SpringerBriefs in Applied Sciences and Technology

SpringerBriefs present concise summaries of cutting-edge research and practical applications across a wide spectrum of fields. Featuring compact volumes of 50 to 125 pages, the series covers a range of content from professional to academic.

Typical publications can be:

- A timely report of state-of-the art methods
- An introduction to or a manual for the application of mathematical or computer techniques
- A bridge between new research results, as published in journal articles
- A snapshot of a hot or emerging topic
- An in-depth case study
- A presentation of core concepts that students must understand in order to make independent contributions

SpringerBriefs are characterized by fast, global electronic dissemination, standard publishing contracts, standardized manuscript preparation and formatting guidelines, and expedited production schedules.

On the one hand, **SpringerBriefs in Applied Sciences and Technology** are devoted to the publication of fundamentals and applications within the different classical engineering disciplines as well as in interdisciplinary fields that recently emerged between these areas. On the other hand, as the boundary separating fundamental research and applied technology is more and more dissolving, this series is particularly open to trans-disciplinary topics between fundamental science and engineering.

Indexed by EI-Compendex, SCOPUS and Springerlink.

José Janiere Silva de Souza
Paulo Cesar Marques de Carvalho

Soiling Effects on Photovoltaic Systems

Theoretical Aspects and Experiences

José Janiere Silva de Souza
Federal Institute of Ceará
Cedro, Brazil

Paulo Cesar Marques de Carvalho
Department of Electrical Engineering
Federal University of Ceará
Fortaleza, Brazil

ISSN 2191-530X ISSN 2191-5318 (electronic)
SpringerBriefs in Applied Sciences and Technology
ISBN 978-3-032-22441-5 ISBN 978-3-032-22442-2 (eBook)
https://doi.org/10.1007/978-3-032-22442-2

This Springer imprint is published by the registered company Springer Nature Switzerland AG
The registered company address is: Gewerbestrasse 11, 6330 Cham, Switzerland

Foreword

The global energy transition toward fossil-free sources is no longer a promise for the future but a pulsing reality that has been reshaping economic and social matrices. At the epicenter of this shift, photovoltaic (PV) technology stands out as one of the most democratic and scalable solutions. However, as solar plants expand into regions of high irradiance—such as the semi-arid zones of the globe—a silent and ubiquitous challenge emerges: **soiling**.

In *Soiling Effects on Photovoltaic Systems: Theoretical Aspects and Experiences*, **José Janiere Silva de Souza** and **Paulo Cesar Marques de Carvalho** offer a fundamental and timely contribution. As Vice-President of the Federal University of Ceará (UFC), it is a matter of great institutional pride to see this research—born and matured within our laboratories—reach the global stage through the publication of this book. The work reflects the academic rigor of our community: **José Janiere** holds a PhD from our Graduate Program in Electrical Engineering, while **Professor Paulo Carvalho**, a CNPq Research Productivity Scholar, represents the excellence that defines UFC as one of Brazil's most prestigious public universities.

This contribution is particularly vital when we consider that Brazil holds the "greenest" electric energy matrix among the world's leading economies. Within this context, the **Brazilian Northeast** emerges as a global powerhouse, with a solar energy potential that far exceeds its current domestic consumption. This surplus is the engine behind our state's most ambitious project: the establishment of a **Low-Emission Hydrogen Hub** at the Port of Pecém (Ceará, Brazil). Through research consortia like **Rede Verdes**, funded by Ceará research funding agency FUNCAP, scientists, government, and industry are working together to transform this natural abundance into sustainable development.

In this landscape, understanding PV efficiency is not merely an academic exercise; it is the cornerstone of the green hydrogen economy. The authors masterfully guide the reader through the essential triad for mitigating losses caused by soiling:

1. **Environmental and Meteorological Monitoring**: Deciphering how wind, precipitation, and humidity dictate deposition and natural cleaning.

2. **Human Activities**: Analyzing how land use—from agriculture to urban expansion—alters the soiling profile.
3. **Cleaning Management**: Establishing maintenance protocols that balance energy gains with the long-term integrity of the modules.

The book brings a state-of-the-art analysis of soiling in PV systems and describes a case study that showcases how empirical observation may translate into engineering strategy. It provides a valuable tool to ensure that solar energy delivers its full potential even in the harshest environments. It is a mandatory reference for students and scientists committed to excellence in the new global energy paradigm.

Universidade Federal do Ceará (UFC)
Fortaleza, Brazil
Diana Azevedo

Competing Interests The authors have no competing interests to declare that are relevant to the content of this manuscript.

Contents

About the Authors

José Janiere Silva de Souza Degree in Industrial Mechatronics Technology from the Federal Institute of Education, Science and Technology of Ceará—Campus Cedro (2017), Master's (2020) and Doctorate (2025) degrees in Electrical Engineering from the Federal University of Ceará (UFC). Currently serves as a Professor of Basic, Technical, and Technological Education at the Federal Institute of Education, Science and Technology of Ceará—Campus Cedro. Teaching courses in the Technical in Electrotechnics, the Technology in Industrial Mechatronics, and the Bachelor's in Electrical Engineering. Academic profile is available at the CNPq Lattes Platform: José Janiere Silva de Souza. Has conducted research at the Alternative Energy Laboratory of UFC.

Paulo Cesar Marques de Carvalho Degree in Electrical Engineering from the Federal University of Ceará (UFC) (1989), Master's degree in Electrical Engineering from the Federal University of Paraíba (1992), and a Doctorate in Electrical Engineering from the University of Paderborn, Germany (1997). Full professor in the Electrical Engineering Department at UFC, working in the electrical and mechanical engineering graduate programs at UFC. Teaching, research, and extension experience in the areas: energy planning, photovoltaic plants, wind power, and biodigesters. Coordinator of the Alternative Energy Laboratory at UFC and CNPq researcher.

Chapter 1
Introduction

1.1 Contextualization

The demand for electricity, driven by industrial and technological growth, has increased substantially. In such context, the International Energy Agency (IEA) projects that electricity will account for more than 50% of final energy consumption in modern economies by 2050 [1]. Thus, to achieve IEA's Net Zero Emissions Scenario, the electricity share in energy demand will need to increase by 4%/year by 2050 [2]. In an energy transition scenario, an important indicator of the electricity sector progress is the share of low-emission technologies (renewables, nuclear, carbon capture and storage, co-firing of ammonia and hydrogen) in the electricity mix [1].

According to the IEA energy transition scenario, to meet the world electricity demand in 2050 based on a stable, affordable, and universal supply—allowing for robust and sustainable economic growth within an economically viable and productive context—total generation will need to grow approximately 2.5 times between 2021 and 2050; to achieve net-zero emissions, investment decisions in new coal-fired power plants must consider emission mitigation measures. Furthermore, the least efficient coal-fired power plants should be decommissioned by 2030, while the remaining ones will need to be modernized by 2040 [3]. In this context, approximately 90% of electricity generation should be supplied by renewable sources, with wind and photovoltaic (PV) power representing about 70% of that total; to achieve this, both sources will need to grow by approximately 11 and 20 times, respectively, between 2021 and 2050 [3].

Historically, solar energy has been used for a variety of applications, such as heating, ventilation, food drying, and water distillation [4]. In the last decades, solar energy for electricity generation on a commercial scale has become a trend in several countries, aiming to replace or reduce the use of fossil fuels [5]. Hence, PV technology is expected to be responsible for approximately 80% of the growth of

J. J. Silva de Souza, P. C. Marques de Carvalho, *Soiling Effects on Photovoltaic Systems*, SpringerBriefs in Applied Sciences and Technology,
https://doi.org/10.1007/978-3-032-22442-2_1

the global capacity from renewable sources, implying an expansion of the number of centralized and decentralized PV systems. It is also projected that PV will surpass wind and hydroelectric generation by 2027 and 2029, respectively, becoming the main renewable energy source in the global electricity matrix [6].

PV modules under operational conditions are significantly influenced by the environmental conditions of the installation site, impacting their electrical performance [7]. One of the critical factors is soiling, resulting from contaminants accumulation on the modules surface [8]. Although soiling impacts were initially neglected by PV producers, in regions with a high occurrence of dust storms or prone to high daily deposition of contaminants, soiling effects assessment is one of the crucial aspects for optimizing PV plants operation [9].

Soiling deposition is influenced by variables such as precipitation, temperature, humidity, wind speed, and inclination and can absorb or disperse the irradiance on the PV modules [10]. Thus, soiling phenomenon varies over time and between different locations; in fact, modules or strings of the same PV plant may present different levels of losses due to soiling. However, the assumption that soiling distribution on the modules surface occurs uniformly is a common premise in most studies reported in the scientific literature [11]. If soiling effects are disregarded, permanent damage to the PV modules may occur, such as the appearance of stains [12]. Soiling is also influenced by different levels and types of atmospheric pollution, which can be caused by human activities carried out in the neighborhood of the PV system site, as well as by dust concentration and/or the occurrence of dust or sandstorms [13]. This phenomenon has been reported as one of the main factors impacting PV performance [14].

Tests carried out near mines in Chitradurga (India) showed that, after 5 days of exposure, a 39.58% reduction of the short-circuit current (I_{sc}) and a 43.18% reduction of the power of a dirty module were observed [15]. After 10 months of deposition under the environmental conditions of the Atacama Desert (Chile), a current reduction of at least 40% (both in I_{sc} and in maximum current—I_{max}) is obtained in p-Si and CdTe modules [16]. When investigating the influence of fouling on m-Si modules installed in Muzarabani (Zimbabwe) for 7 months, average losses ranging from 0.29%/day (8.84%/month) to 0.48%/day (14.48%/month) were reported [17]. Examining the soiling effect on p-Si modules used in streetlights installed in different areas—coastal, industrial, urban, and rural—of Kupang (Indonesia), it was observed that modules power was reduced between 9.14% and 13.94%, with I_{sc} being the most affected parameter (reductions of up to 19%). The coastal zone presented the greatest losses (between 11.27% and 13.94%), followed by urban areas (12.02%–13.53%), industrial areas (9.82%–10.37%), and rural areas (9.13%–9.75%) [18].

Although there has been a growing interest in studying this topic in recent decades [19], the effects of soiling on solar systems were first observed in the 1940s in solar thermal collectors, as reported in [20]. Regarding PV systems, research has been carried out to demonstrate soiling effects on different PV technologies, especially in arid and semi-arid regions, so that interest in this topic tends to grow with the increase in the implementation of PV plants. In addition to impacting generation performance, soiling increases operation and maintenance (O&M) costs and, consequently, the financial risk of commercial-scale PV plants [21]. Depending on the PV plant characteristics, soiling mitigation can represent up to 70% of O&M costs [22]. Such share is due to the demand for labor for routine interventions, increasing expenses and reducing projects profit margin; in certain locations, cleaning interventions may prove economically unfeasible [12].

Commercial-scale PV generation is the most competitive form of electricity generation in many parts of the world, but it faces significant challenges due to the limited availability of large areas with ideal solar resources [23] and the effects of climate change [24]. Alternatively, PV systems in urban environments have been widely adopted in applications such as parking canopies [25], electric vehicle charging stations [26], and Building Integrated Photovoltaic (BIPV) systems [27]. However, urban applications present additional challenges: low wind speeds that hinder heat dissipation from modules, shadows from buildings and pollution-related problems [5, 27].

After installation, one of the main strategies to improve PV efficiency is the use of soiling mitigation methods. However, the adoption of inappropriate cleaning strategies, which disregard local climate conditions, labor costs, electricity prices, PV technology, and natural degradation, can compromise the financial viability of PV systems [28, 29]. Despite numerous technological solutions under investigation (drones, cleaning robots, self-cleaning coatings, and automated/semi-automated cleaning systems) [30], manual cleaning remains one of the most widely applied solutions for PV systems in urban environments. This approach is particularly relevant in arid and semi-arid regions, as low-volume and irregular precipitations can remove larger particles, but its effectiveness in eliminating smaller particles is limited [31].

Global soiling losses were estimated at between 3% and 4% in 2018, representing a financial loss of between three and five billion euros; even with the adoption of optimized schedules, soiling losses were in the order of 4%–5% in 2023. One of the factors that justifies this process is the increase in the number of new PV installations in regions with high exposure to soiling, such as China, India, and MENA (Middle East and North Africa) [24]. However, this estimate does not consider the additional cleaning costs resulting from the adoption of inadequate schedules, and the cleaning of PV systems installed on rooftops, which cost between three and eight times more to clean than ground-based systems. However, the IEA report "Soiling Losses—Impact on the Performance of Photovoltaic Power Plants" [32] highlights that the increase in the number of new PV systems on rooftops tends to aggravate the global soiling impacts. The effects of climate change should also not be underestimated, as the increase in the global temperature causes greater aridity in soil, raising the risk of forest fires and the particulate matter (PM) suspended in the air [24]. Furthermore, in countries such as Brazil, where most of the population lives in urban areas, the interest of residential and commercial consumers in reducing electricity costs has driven a significant increase in the distributed PV capacity, which already exceeds that of centralized PV plants [33].

1.2 Overview of Chapters

Chapter 1—Introduction: The global landscape of electricity generation is discussed, highlighting data from the International Energy Agency (IEA) on the growing future demand and the importance of low-carbon technologies to meet this need. Sequentially, historical uses of solar energy by humanity are mentioned, culminating in the introduction of PV systems as a modern and sustainable alternative in the current energy context. Subsequently, factors affecting PV performance are presented, with an emphasis on the process of soiling. Soiling implications on generation efficiency are discussed, considering local environmental conditions and the additional operation and maintenance costs associated with mitigation strategies.

Chapter 2—Theoretical Basis: Theoretical aspects related to PV systems and soiling process are addressed. Initially, concepts about solar resources and PV technologies are presented, highlighting the solar irradiance distribution and its components: direct, diffuse, and albedo. Furthermore, the PV effect and the main factors that influence PV cells performance are discussed. Subsequently, the physical and chemical composition of soiling, the stages of the soiling cycle, the mitigation methods, and the metrics used to evaluate the soiling impact on PV electrical performance are commented on.

Chapter 3—State of the Art—Perspectives: Perspectives observed in studies on soiling are presented. Aspects related to the main PV technologies used, as well as the most employed methods for assessing soiling effects in different parts of the world are explored. In this context, considerations about natural cleaning processes and the limitations associated with certain cleaning methods are commented. Additionally, the main effects of soiling on PV production and the most observed physical and chemical characteristics of soiling composition are discussed.

Chapter 4—Case Study—Experimental Procedures: The methodological procedures adopted in a case study conducted in the city of Fortaleza, Brazil are presented. Initially, the PV plant and the data-acquisition system used are described. Environmental and operational data, including irradiance, wind speed, humidity, and module temperatures, are monitored using advanced sensors calibrated for greater accuracy. Additionally, the manual cleaning procedures applied, the materials used, and the considerations adopted for defining the evaluation metrics are commented. The procedures are divided into three phases: no cleaning, complete cleaning, and selective cleaning of string 2 (ST2) modules.

Chapter 5—Case Study—Experimental Results: The effects of soiling and manual cleaning on the PV electricity generation are analyzed. Soiling metrics are assessed, comparing the generation of clean and dirty modules, with an emphasis on the effects of cleaning interventions over time. The relationship between cleaning frequency, precipitation, and generation is addressed, demonstrating soiling tendency to increase energy losses. The analysis compares performance metrics of clean and dirty modules, with a focus on the gains observed in the evaluated metrics, particularly of modules that receive more cleaning interventions.

Chapter 6—Final Considerations: A synthesis of soiling impacts on PV systems and the importance of adopting mitigation measures is proposed. The results of a case study are discussed, highlighting how soiling affects the generation, particularly during the local dry period, and how cleaning interventions can be effective in improving module performance. Additionally, the research emphasizes the importance of cleaning frequency and its relationship with rainfall, demonstrating soiling's tendency to increase energy losses.

References

1. International energy agency—IEA: Electricity sector, Paris (2022), https://www.iea.org/reports/electricity-sector. Accessed 19 June 2025
2. International Energy Agency—IEA: Electricity sector, Paris (2025), https://www.iea.org/energy-system/electricity/electrification. Accessed 19 June 2025
3. International Energy Agency—IEA: Net Zero by 2050, Paris (2021), https://www.iea.org/reports/net-zero-by-2050. Accessed 08 Nov 2025
4. P. Carvalho, *O Uso da Energia: 4 Milhões a.C. Aos Tempos Atuais* (Lisbon International Press, Brazil, 2022)
5. H.A. Kazem, M.T. Chaichan, A.H.A. Al-Waeli, K. Sopian, A review of dust accumulation and cleaning methods for solar photovoltaic systems. J. Clean. Prod. **276**, 123187 (2020). https://doi.org/10.1016/j.jclepro.2020.123187
6. International Energy Agency—IEA: Solar PV, Paris (2025), https://www.iea.org/energy-system/renewables/solar-pv. Accessed 08 Nov 2025
7. L.C. de Lima, L. de Araújo Ferreira, F.H.B. de Lima Morais, Performance analysis of a grid connected photovoltaic system in northeastern Brazil. Energy Sustain. Dev. **37**, 79–85 (2017). https://doi.org/10.1016/j.esd.2017.01.004
8. W.J. Jamil, H. Abdul Rahman, S. Shaari, Z. Salam, Performance degradation of photovoltaic power system: Review on mitigation methods. Renew. Sust. Energ. Rev. **67**, 876–891 (2017). https://doi.org/10.1016/j.rser.2016.09.072
9. S. L., N. Kumaar, H.P. Waldl, P.K. Das, K. Ramanathan, K. Balaraman, I. Mitra, Impact of soiling on energy yield of solar PV power plant and developing soiling correction factor for solar PV power forecasting. European Journal of Energy Research **1**(2), 21–29 (2021). https://doi.org/10.24018/ejenergy.2021.1.2.7
10. S. Nurjanah, T. Dewi, Rusdianasari, Dusting and soiling effect on PV panel performance: Case study open-pit mining in South Sumatra, Indonesia, in *2021 International Conference on Electrical and Information Technology (IEIT)*, (IEEE, 2021). https://doi.org/10.1109/ieit53149.2021.9587351
11. J.G. Bessa, L. Micheli, F. Almonacid, E.F. Fernández, Monitoring photovoltaic soiling: Assessment, challenges, and perspectives of current and potential strategies. IScience **24**(3), 102165 (2021). https://doi.org/10.1016/j.isci.2021.102165
12. M. Hammoud, B. Shokr, A. Assi, J. Hallal, P. Khoury, Effect of dust cleaning on the enhancement of the power generation of a coastal PV-power plant at Zahrani Lebanon. Sol. Energy **184**, 195–201 (2019). https://doi.org/10.1016/j.solener.2019.04.005
13. T. Alkharusi, G. Russo, A. Oyeniran, C. Pandey, A. Buonomano, A. Palombo, C.N. Markides, Quantifying the influence of soiling on PV module energy production based on a coupled opto-thermal-electrical model: An international assessment in different climatic conditions. Renew. Energy, 124119 (2025). https://doi.org/10.1016/j.renene.2025.124119

14. Z. Song, J. Liu, H. Yang, Air pollution and soiling implications for solar photovoltaic power generation: A comprehensive review. Appl. Energy **298**, 117247 (2021). https://doi.org/10.1016/j.apenergy.2021.117247
15. A.K. Tripathi, M. Aruna, C.S.N. Murthy, Performance evaluation of PV panel under dusty condition. Int. J. Renewable Energy Dev. **6**(3), 225 (2017). https://doi.org/10.14710/ijred.6.3.225-233
16. P. Ferrada, D. Olivares, V. del Campo, A. Marzo, F. Araya, E. Cabrera, J. Llanos, J. Correa-Puerta, C. Portillo, D. Román Silva, M. Trigo-Gonzalez, J. Alonso-Montesinos, G. López, J. Polo, F.J. Batlles, E. Fuentealba, Physicochemical characterization of soiling from photovoltaic facilities in arid locations in the Atacama Desert. Sol. Energy **187**, 47–56 (2019). https://doi.org/10.1016/j.solener.2019.05.034
17. K. Chiteka, R. Arora, S.N. Sridhara, C.C. Enweremadu, A novel approach to solar PV cleaning frequency optimization for soiling mitigation. Sci. Afr. **8**, e00459 (2020). https://doi.org/10.1016/j.sciaf.2020.e00459
18. J. Tanesab, R. Sinaga, J. Mauta, A. Amheka, E. Hattu, F. Nikson Liem, Temporary performance degradation of photovoltaic street light in Kupang city, Nusa Tenggara Timur province, Indonesia. E3S Web Conf. **190**, 00018 (2020). https://doi.org/10.1051/e3sconf/202019000018
19. J.J.S. Souza, P.C.M. Carvalho, G.C. Barroso, Analysis of the characteristics and effects of soiling natural accumulation on 2 photovoltaic systems: A systematic review of the literature. J. Solar Energy Eng., 1–99 (2022). https://doi.org/10.1115/1.4056453
20. H. Hottel, B. Woertz, Performance of flat-plate solar-heat collectors. Trans. ASME (Am. Soc. Mech. Eng.); (United States), 64 (1942), https://www.osti.gov/biblio/5052689
21. R. Nahar Myyas, M. Al-Dabbasa, M. Tostado-Véliz, F. Jurado, A novel solar panel cleaning mechanism to improve performance and harvesting rainwater. Sol. Energy **237**, 19–28 (2022). https://doi.org/10.1016/j.solener.2022.03.068
22. V. Gonçalves, Aumento na produtividade com a utilização de robô de limpeza. Canal Solar (2024), https://canalsolar.com.br/aumento-na-produtividade-com-a-utilizacao-de-robo-de-limpeza/
23. International Energy Agency—IEA: Solar PV, Paris (2022), https://www.iea.org/energy-system/renewables/solar-pv. Accessed 19 June 2025
24. K. Ilse, L. Micheli, B.W. Figgis, K. Lange, D. Daßler, H. Hanifi, F. Wolfertstetter, V. Naumann, C. Hagendorf, R. Gottschalg, J. Bagdahn, Techno-economic assessment of soiling losses and mitigation strategies for solar power generation. Joule **3**(10), 2303–2321 (2019). https://doi.org/10.1016/j.joule.2019.08.019
25. R. Maier, L. Lütz, S. Risch, F. Kullmann, J. Weinand, D. Stolten, Potential of floating, parking, and Agri photovoltaics in Germany. Renew. Sust. Energ. Rev. **200**, 114500 (2024). https://doi.org/10.1016/j.rser.2024.114500
26. M. Yao, D. Da, X. Lu, Y. Wang, A review of capacity allocation and control strategies for electric vehicle charging stations with integrated photovoltaic and energy storage systems. World Electric Vehicle J. **15**(3), 101 (2024). https://doi.org/10.3390/wevj15030101
27. W. Wang, H. Yang, C. Xiang, Green roofs and facades with integrated photovoltaic system for zero energy eco-friendly building—A review. Sustain Energy Technol Assess **60**, 103426 (2023). https://doi.org/10.1016/j.seta.2023.103426
28. H. Yazdani, M. Yaghoubi, Dust deposition effect on photovoltaic modules performance and optimization of cleaning period: A combined experimental–numerical study. Sustain Energy Technol Assess **51**, 101946 (2022). https://doi.org/10.1016/j.seta.2021.101946
29. M. Mehdi, R. Conceição, N. Ammari, A. Alami Merrouni, J. González-Aguilar, M. Dahmani, Modeling, assessment and characterization of soiling on PV technologies. Toward a better understanding of the relation between dust deposition and performance losses. Sustain Energy Technol Assess **71**, 104023 (2024). https://doi.org/10.1016/j.seta.2024.104023
30. H.M. Khalid, Z. Rafique, S.M. Muyeen, A. Raqeeb, Z. Said, R. Saidur, K. Sopian, Dust accumulation and aggregation on PV panels: An integrated survey on impacts, mathematical models, cleaning mechanisms, and possible sustainable solution. Sol. Energy **251**, 261–285 (2023). https://doi.org/10.1016/j.solener.2023.01.010

31. T. Alkharusi, G. Huang, C.N. Markides, Characterisation of soiling on glass surfaces and their impact on optical and solar photovoltaic performance. Renew. Energy, 119422 (2023). https://doi.org/10.1016/j.renene.2023.119422
32. International Energy Agency Photovoltaic Power Systems Programme (IEA PVPS): Soiling losses—impact on the performance of photovoltaic power plants (2022), https://iea-pvps.org/key-topics/soiling-losses-impact-on-the-performance-of-photovoltaic-power-plants/. Accessed 19 June 2025
33. ASSOCIAÇÃO BRASILEIRA DE ENERGIA SOLAR—ABSOLAR: Energia solar fotovoltaica no Brasil: Infográfico ABSOLAR (2025), https://www.absolar.org.br. https://www.absolar.org.br/mercado/infografico/. Accessed 19 June 2025

Chapter 2
Theoretical Basis

2.1 Solar Resource

Solar irradiation is the flow of energy emitted by the Sun that propagates in the form of waves. Due to temperature gradients and the various emission and absorption lines on the Sun's surface, solar electromagnetic irradiation, also called solar spectrum, is similar to that of a black body with a temperature of around 5,800 K, which is the approximate temperature of the Sun's surface (photosphere). The solar spectrum can be classified into bands according to wavelength or frequency; most of the visible range (400–780 nm) is emitted from the photosphere, the other bands are infrared (700–4000 nm) and ultraviolet (200–400 nm). The average annual density of solar irradiance measured in a plane perpendicular to the direction of propagation of the Sun's rays at the top of the atmosphere is called the solar constant and corresponds to the value of 1,367 W/m^2. When the Earth is closest to the Sun (perihelion—January 3), extraterrestrial irradiance corresponds to 1,400 W/m^2, and when it is furthest away (aphelion—July 4), the value is approximately 1,330 W/m^2. The total power available by the Sun to Earth is estimated at 174,000 TW; however, due to absorption, reflection, and diffusion in the atmosphere and at the Earth's surface, approximately 94,000 TW reaches the Earth's surface [1, 2].

Due to interactions with the atmosphere and the environment, global irradiance can be decomposed into three components: direct, diffuse, and reflected irradiance. Direct irradiance is the portion that comes directly from the Sun and passes through the atmosphere until reaching the receiver. Diffuse irradiance originates from interactions with elements present in the atmosphere, such as clouds, having no specific propagation direction. Reflected irradiance, or albedo, originates from irradiance reflection by surfaces surrounding the receiver [3].

J. J. Silva de Souza, P. C. Marques de Carvalho, *Soiling Effects on Photovoltaic Systems*, SpringerBriefs in Applied Sciences and Technology,
https://doi.org/10.1007/978-3-032-22442-2_2

2.2 Photovoltaic Technology

2.2.1 Photovoltaic Effect

Electrons in an isolated atom have discrete energy levels. When atoms approach each other, the individual electron energy levels change, causing the energy levels to group into bands. The electrons in the valence orbit are the only ones that can interact with other atoms and, when grouped together, form a valence band; some electrons in the valence band can acquire enough energy to move into a conduction band. The energy difference between the valence band and the innermost shell of the conduction band is called band gap or simply gap [2, 4]. Semiconductor materials have a valence band filled with electrons and a conduction band that remains unchanged at 0 K, meaning they behave like insulators. Due to these energy bands, semiconductors increase their conductivity as the temperature rises, as thermal excitation promotes electrons from the valence to the conduction band. At temperatures above 0 K, electrons are found in the conduction band; the same number of holes is counted in the valence band [1].

The property that enables PV conversion is the creation of electron-hole pairs from the incidence of photons with energy greater than the band gap energy (E_g) of the material. However, to harness the electricity, an electric field is required to separate the carriers; this is possible through a PN junction [1]. To form a PN junction, impurities are added in a controlled manner into a pure (intrinsic) semiconductor through a doping process. When pentavalent atoms, Group 15 of the periodic table, such as phosphorus (P), are introduced into a crystal lattice, an excess electron is weakly bound to its parent atom. A semiconductor doped with pentavalent impurities is called an N-type (extrinsic) semiconductor. On the other hand, if the dopant is a trivalent atom, Group 13 of the periodic table, such as boron (B), there is an absence of electrons (holes) in the crystal lattice. A semiconductor doped with trivalent impurities is called a P-type (extrinsic) semiconductor [5, 6].

At the interaction interface between P and N materials, the excess electrons in the N-type semiconductor diffuse into the P-type semiconductor, creating positive ions due to electron loss. As they diffuse into the P-type semiconductor, the electrons occupy the existing holes, generating negative ions. This movement is interrupted by the generation of an electric field, and equilibrium is achieved with the emergence of the depletion zone, where positive and negative charges are present, representing a potential barrier. In laboratory conditions, PN junction formation occurs through the diffusion of gaseous P compounds on the front face of a crystalline Si wafer in high-temperature furnaces (870 °C), in 15–30 min, allowing the formation of an N-type semiconductor approximately 0.50 μm thick [4]; the P-type material is formed by introducing a small amount of B into molten Si [5]. Sequentially, an anti-reflection coating and electrical contacts are inserted on the cell front side; an electrical contact is added on the back side [1].

The light that strikes a PV cell can be reflected, absorbed, or transmitted. If the photon is absorbed by an electron, its energy increases; photons with high frequencies, or short wavelengths, are the most energetic. If the absorbed photon energy is greater than the semiconductor band gap, this electron jumps into the conduction band and circulates freely, leaving a hole in the valence band. Hence, an electron-hole pair is created, and the photon excess energy is converted into heat; even in this case, only a single electron-hole pair is generated, which is one of the reasons for the low PV conversion efficiency [4]. If the electron-hole pair is generated in or near the depletion zone, it is directed by the electric field in the PN junction, generating a current if a load is connected, in the direction from the N to the P region [1, 2, 7]. If the photon energy is lower than the band gap, the electron does not obtain enough energy to pass to the conduction band.

2.2.2 Spectral Response of Photovoltaic Technologies

Electrical characteristics of PV cells are influenced by the manufacturing process, materials used, internal resistances, and absorption selectivity, as well as environmental factors such as irradiance, ambient temperature, and soiling. Absorption selectivity of the solar spectrum relates to the energy level required for a photon to be absorbed by a PV cell, generating an electron-hole pair [8]. In the case of crystalline Si cells, PV cells operate in a range corresponding to visible light and near-visible infrared. Due to the lack of coincidence between the maximum energy of each wavelength in the solar spectrum and the PV cells' maximum spectral response, not all incident irradiation is utilized.

Ideally, PV cell spectral response would increase linearly with wavelength; however, in real cells, such as monocrystalline silicon (m-Si), the maximum response occurs around 800 nm, decreasing from that point. The distinct spectral response of each technology is due to optical losses caused by absorption and reflection by the module coating (glass), electron-hole recombination processes, defects, impurities, and differences in semiconductor doping. In the case of multijunction cells, each junction presents a distinct spectral response, aiming to maximize the use of the solar spectrum. Thus, the spectral response is the combination of the spectral response of each individual junction [9].

2.2.3 Factors Influencing Photovoltaic Cell Performance

PV modules are classified under Standard Test Conditions (STC), which do not necessarily correspond to the environmental conditions of a particular locality; in operational conditions, PV modules are susceptible to the effects of high temperatures, air pollution, shading, and soiling [10]. Shading on PV modules can occur partially or totally and is caused by several factors, such as soiling accumulation, bird droppings, clouds, trees, buildings or chimneys, as well as physical damage, such as breakage of the modules glass cover; this phenomenon significantly reduces

PV performance and is considered one of the most common and impactful limitations of solar plants [11, 12]. PV modules electrical characterization includes power-voltage (P-V) and current-voltage (I-V) curves, which graphically represent modules operation under certain conditions; these curves provide information that allows the adoption of strategies for the operation near the Maximum Power Point (MPP). However, PV modules electrical characteristics under shading are modified; multiple peaks are formed at the P-V curve, causing false tracking of the MPP, resulting in power loss [12].

PV cell temperature (T_c) depends on the module design, materials used, and, primarily, environmental conditions, and is considered a critical parameter for characterizing the performance of these devices [13]. Therefore, accurate estimation of T_c is essential for PV plants design and operation, as assessment errors can result in over or underestimation of power output, directly impacting PV plants performance and financial return [14]. In general, a reduction of the open-circuit voltage (V_{oc}) and fill factor (FF) is observed with an increasing temperature, while I_{sc} tends to increase slightly, leading to a decrease of modules efficiency (η); V_{oc} reduction is less pronounced in materials with a larger bandgap [10]. For crystalline Si modules, an efficiency decrease of approximately 0.5% for each 1 °C increase of T_c is observed [15].

2.3 Soiling

Soiling can be defined as a layer of dust, snow, or any contaminant material accumulating on PV modules surface, causing the reduction of glass transmittance and, consequently, reduction of PV electrical performance [16, 17]. A module with soiling may be subject to partial shading conditions, depending on the distribution (uniform or nonuniform) and the thickness of the soiling layer. Typically, nonuniform deposition is observed in PV modules with frames; in such a case, there is a greater propensity for soiling accumulation at the lower edge due to the displacement of the particles by precipitation, especially in modules with small slope angles. Nonuniform deposition, which may be related to the presence of animal droppings and the proliferation of fungi, moss, and algae, compromises PV electrical performance and can cause the emergence of hot points [18]; examples of different types of soiling deposition are shown in Fig. 2.1.

PV electrical performance reduction can reach up to 1% per day, especially in desert regions [19]. Soiling process is reversible, but it is a global, inevitable, and challenging feature. Studies on soiling impacts are still limited, as they depend on specific local weather conditions; additionally, the influence of anthropogenic activities near PV plants should be considered as they represent an important source of soiling particles and are commonly neglected during the PV system implementation phase. In this sense, evaluations on different mitigation processes should be systematically analyzed, aiming to maximize PV production, especially in large urban environments. According to the literature, natural soiling mitigation by precipitation

Fig. 2.1 Examples of different types of soiling deposition

is not sufficient for the complete cleaning of PV modules installed in this type of environment [20]. Soiling tends to remain a global challenge due to climate change, increased global temperature and the occurrence of prolonged drought periods. Hence, studies focused on modeling, adaptation, and soiling mitigation strategies are essential to maximize PV plants generation and financial return [21].

2.3.1 Soiling Composition

The characterization of the soiling composition and distribution through PV modules surface is a crucial factor for understanding interactions between particles and modules, determining their effects on electrical/thermal performance and the choice of mitigation methods appropriate to the particularities of each site. Soiling particle size depends on the source and the mechanisms of generation and dispersion; chemical characteristics are also specific to each location, as they are subject to production mechanisms and generating sources [22]. Therefore, soiling under different climatic/environmental conditions should be characterized and measured for a comprehensive understanding; such an aspect represents a barrier to establishing a universal pattern for PV generation reduction due to soiling [23].

The loading of air-suspended particles is mainly related to wind erosion; after the start of particles movement, transport mechanisms depend on their size [24]. Larger particles are mainly affected by inertial and gravitational effects, while smaller particles are influenced by interparticle interactions. As a result, thin particles (<1 μm) tend to accumulate faster than thick particles (>5 μm) [23]. Particles present in the air come from natural processes or derive from anthropogenic activities; generally, the burning of fuels, mechanical processes, grinding, mineral processing and soiling resuspension, for example, result in the emission of a set of chemical elements associated with each process [25].

2.3.2 *Soiling Deposition Process*

To understand and eventually control soiling, it is necessary to know the process stages, shown in Fig. 2.2 [26]. Particle deposition mechanisms can be classified as: sedimentation by gravity, inertial or turbulent deposition and Brownian movement. The performance of these mechanisms depends on the transported particle size and wind speed: 1 μm-size particles spread in the random space and have greater propensity to remain in the wind flow (Brownian movement); particles with size ≥10 μm are predominantly deposited by inertia; particles about 100 μm have gravitational sedimentation as a predominant mechanism. Sedimentation severity is influenced by particle characteristics such as size, density, and morphology. Inertial deposition, in turn, depends on the speed and turbulence of the wind. In the case of PV modules, the sedimentation process can be mitigated by their inclination or by inserting a soiling protection coverage when not in operation. The influence of inertial deposition can be reduced by placing the PV plant in a locality with a standard air flow or above ground level, where the wind tends to be faster [27].

During the transport step, particles are removed from air continuously by dry and/or moist deposition processes, with a temporal and spatial dependence on these mechanisms. Dry deposition results from gravitational sedimentation and mixture of suspended particles that deposit on the PV modules, while wet deposition is characterized by the elimination of suspended particles due to precipitation or snow [28]. In the initial adhesion stage of the modules surface, several forces act between particles and surface, such as hair forces, van der Waals, electrostatic and gravitational. Under wet conditions, capillary forces represent 98% of the forces acting between particles and surface, while van der Waals forces correspond to 2%. Electrostatic forces are considered negligible, as humidity acts in eliminating their effects; gravitational forces are irrelevant to particles less than 500 μm [29].

Capillary forces dominate when moisture interacts with particles and surface through the formation of water bridges [30]; their performance is influenced by the particle diameter (especially effective for particles >10 μm), moisture (levels between 30 and 40% are sufficient), surface geometry, and contact angles [29, 30]. On the other hand, hair forces are reduced on hydrophobic surfaces, with water

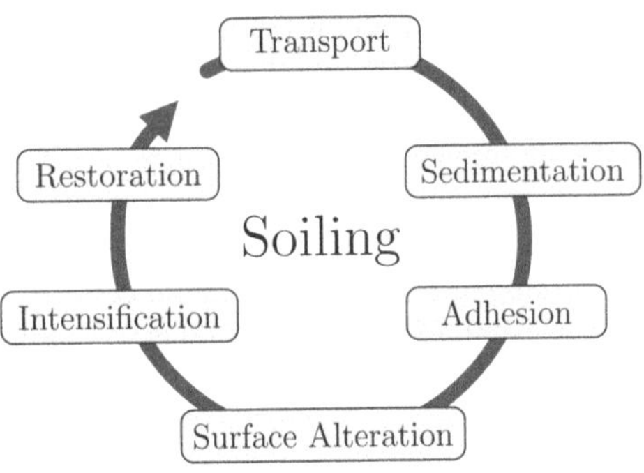

Fig. 2.2 Soiling process stages

repellency and high roughness [31]. Van der Waals' strength results from attraction or repulsion between atoms and molecules [29]; the force has a short range and tends to zero when surfaces initially in contact are separated. Thus, there is a dependence on the contact area between particles and surface, material properties, and the presence of moisture [31].

Suspended air particles can acquire electric charge by collisions and other means, even when the modules surface is not charged, giving rise to electrostatic forces [29]. These forces are influenced by electrical fields, electrical properties of charge and surface, separation distance, and charge state of particles; usually, electrostatic forces are not dominant for distances less than 20 nm [31]. PV modules in operation are subject to natural cleaning mechanisms, but the particles adherence process to their surface can be intensified by precipitations of low intensity and frequency, high humidity and dew formation, resulting in cementation, particle agglomeration, and capillary aging [31, 32].

In the presence of water or in conditions with high moisture, the soluble part of soiling tends to form microscopic droplets of saline solutions involving the non-soluble part [32]. During the drying process, the precipitated saline material causes the formation of solid bridges between the non-soluble material and the PV modules surface, acting as cement; consequently, particles cannot be removed by bearing or sliding. If this process occurs periodically, without cleaning interventions, cemented particle layers accumulate, the cleaning process (manual or natural) becomes more difficult and, in extreme cases, PV modules properties are permanently altered. Particle agglomeration is caused by rearrangement, agglomeration, and particle compaction during dew events. The formation of water drops on the module surface causes the suspension of small particles; as the drop dries, particle agglutination occurs at a smaller volume until they come into contact again with the module surface. Thus, small particles allocate among the spaces formed by larger particles, uniting them after water evaporation and increasing van der Waals forces [31].

Capillary aging is caused by the flattening of soiling particles against the PV modules surface due to hair forces when liquid bridges between particles and the surface are drying. As a result, there is an increase in contact area and intensification of adhesion forces, such as van der Waals forces. Additionally, this adhesion mechanism can occur between particles, promoting agglomeration and cementation and hindering wind removal [31]. Due to soiling adhesion mechanisms, modules surface may change their original characteristics, resulting in greater particle accumulation [26]. Characteristics such as apparent texture and additional coating influence the soiling settlement rate [33]; therefore, mitigation methods, if well employed, may prevent damage of modules integrity, as soiling spots may induce PV cells overheating and cause a short-circuit [34].

2.4 Soiling Mitigation Methods

Initially, soiling should be properly monitored to determine the best mitigation method since no standard strategy is valid globally due to the influence of local characteristics [35]. In general, mitigation strategies are divided into two categories: cleaning and preventive actions.

The adoption of an optimized soiling mitigation strategy offers the following advantages: (1) increased PV plant revenues due to the Capacity Factor (CF) increment and (2) improvement of financial margins, encouraging new investments. Although daily PV plant performance losses vary according to the locality, several experimental studies show that these losses accumulate over time. Therefore, the choice of the cleaning method, periodicity, efficiency, and associated costs depends on a previous case study. In addition, aspects such as PV plant size, water availability, weather parameters, and soiling characteristics should be considered [36].

Soiling deposition rate and cleaning frequency can change along the seasons in a year or from year to year, due to the variability of weather and environmental conditions. Additionally, periodic human activities, such as harvests, and extraordinary events, such as sandstorms, should be considered. Understanding these seasonal processes and, eventually, their distribution in a year is fundamental to choosing the ideal mitigation strategy for a site [35]. In the short term, the adoption of cleaning schedules can reduce the effects of contaminant deposition on PV power, especially in the case of highly polluted urban areas.

2.4.1 Natural Cleaning

Under certain conditions, natural cleaning caused by heavy rainfall or snowmelt may be sufficient to restore PV performance impacted by soiling [37]. Precipitations play an important role in mitigating soiling, especially for PV systems with no cleaning schedule due to the lack of water or skilled labor [38]. However, low-intensity precipitations can contribute to a greater contamination deposition, accelerating the soiling process [39]. Wind can also help the natural cleaning of PV modules, especially moderate wind carrying no dust [36]. PV modules surface roughness reduces the adhesion of soiling particles at all humidity levels: in dry conditions, there is a reduction of the surface-particle contact area; in humid conditions, liquid bridges form between the surface and the particles. However, considering natural mitigation procedures, eliminating the occurrence of condensation on the surface of dirty modules is difficult, as this process can occur through the formation of microscopic droplets, even on hydrophobic surfaces [40].

2.4.2 Artificial Cleaning

Manual mechanical cleaning is like cleaning windows in buildings with soft bristle brushes and may be more efficient than natural cleaning or non-manual mechanical cleaning with tap water. However, direct interaction with the surface can cause physical damage to the modules, such as abrasions and cracks, due to uneven movements/pressure or improper handling of the equipment. Furthermore, specialized labor is required, resulting in higher costs compared to other passive methods [36]. The risk of personnel accidents, difficult movements, and water waste are other disadvantages of the manual cleaning method [41]. Non-manual mechanical cleaning uses semi-automatic devices that require worker intervention/supervision, or fully automatic devices that, supported by sensors, move brushes across the modules surface, using water or not. Systems based on dry non-manual mechanical cleaning can be a viable solution for areas with limited access to water; however, direct contact of the brushes with the modules surface can cause physical damage. Another limiting factor is the additional energy consumption, the need for maintenance of mechanical parts, and the high complexity of controlling robots, motors, or scanners [36].

Wet non-manual mechanical systems act similarly to precipitations, channeling water to the PV modules. However, this method requires water availability and the use of pumps to increase water pressure; in some cases, cleaning agents can be added to aid in the particle removal process. Other limiting factors are energy consumption by pumps, clogging/rupturing of hoses, and the possibility of thermal shocks on the modules surface since this type of cleaning is often activated when the sun is very intense. Another mitigation method is the use of electrostatic screens (Electro-Dynamic Displays—E-DS), based on the use of a high-voltage source to produce an electric field on a screen attached to the modules surface, aiming to charge the dust particles, facilitating their removal [36]. The main features include rapid removal and ease of control; the disadvantages are degradation of the E-DS screen due to ultraviolet radiation, the need for a high-voltage source to produce the electric field, and limited removal of dry and small particles.

2.4.3 Preventive Actions

The tilt angle directly controls the amount of soiling deposition, as gravity tends to pull particles toward the bottom or out of the modules as they approach the vertical [22]; greater tilt angles hinder the deposition of soiling particles, facilitating their removal either by gravitational forces or by the action of natural cleaning agents (rainfall and wind, for example) [42]. PV modules self-cleaning methods are used to reduce labor costs in areas that are difficult to access [43]. The use of coatings involves applying hydrophobic solutions in thin layers to the modules surface, acting as a barrier against soiling cementation; when water precipitates, the droplets roll across the surface, carrying away the soiling particles already deposited. Due to ultraviolet irradiation, the method's effectiveness can be limited by coating degradation. Furthermore, hydrophobic self-cleaning coatings require the presence of water to remove soiling, limiting their use in dry environments [36].

Regarding the condensation process on surfaces with hydrophobic coatings, laboratory experiments have consistently shown a reduction of soiling particle adhesion at different relative humidity levels [40]. However, hydrophobic coatings have not necessarily been able to reduce soiling accumulation in field experiments. Therefore, their use does not eliminate the need for water to optimize cleaning results, one of the reasons for the limited use of coatings [44].

2.5 Soiling Effects Estimation in PV Plants

Models are found in the literature correlating environmental variables (wind, humidity, temperature) and soiling data; some models include modules tilt angle as a variable to model the soiling effect. Another widespread approach is the adoption of Artificial Neural Networks or Computational Fluid Dynamics [44]. However, the number of factors that influence the soiling process tends to restrict such models to a specific location.

For greater effectiveness in modeling soiling effects, studies should be based on data from at least 1 year of practical observations, aiming to incorporate seasonal characteristics of the site. The use of data from a longer observation period allows for a more assertive analysis, detecting seasonal similarities of soiling rates during different years and verifying their dependence on atmospheric parameters [44]; hence, continuous and regular monitoring of PV plant operations should be established. Various automated approaches have been proposed to reduce costs and human intervention: use of soiling stations, optical sensors, image analysis sensors, and soiling extraction algorithms. However, implementing some of these solutions requires significant financial investment in the installation and/or acquisition stages. On the other hand, the extraction of soiling effects, in their simplest form, can be estimated using performance data or environmental parameters, even in plants where no module's cleaning strategy is implemented [35].

The variability and interaction of environmental parameters are fundamental to the process of soiling deposition and removal [35]. Therefore, by understanding the correlations with these parameters, it is possible to estimate the soiling effects without the need for PV plant performance analysis, estimating soiling losses in advance during project design, cost assessment, and installation site selection. Many characteristics influence soiling impacts, such as particle composition, size, and shape, as well as environmental parameters such as humidity, temperature, and wind [35]. Hence, soiling deposition results from the interaction of several variables and cannot be directly determined by a single parameter [44].

In this context, one of the main challenges in studying soiling through the calculation of performance metrics is identifying and eliminating issues unrelated to the soiling process, such as module and inverter degradation and failures. Regarding the use of evaluation metrics, the need to standardize symbols and nomenclatures stands out to ensure their proper use [44]. Different soiling assessment metrics are found in the literature; however, it is initially necessary to make parameter corrections for STC, so that PV plants performance can be compared regardless of their geographic location, positioning, tilt, and nominal power [45]. Equations (2.1) and (2.2) are used to correct I_{sc} and maximum power (P_{Max}); the subscript STC is used to describe the electrical characteristics of the modules reported in the manufacturer catalog.

$$I_{sc_{Norm}} = I_{sc_{STC}} \cdot \left[1+\alpha\left(T_c - T_{STC}\right)\right] \cdot \frac{G_{POA}}{G_{STC}} \tag{2.1}$$

$$P_{Max_{Norm}} = P_{max} \cdot \left[1+\gamma\left(T_c - T_{STC}\right)\right] \cdot \frac{G_{POA}}{G_{STC}} \tag{2.2}$$

The parameters $I_{sc_{Norm}}$ and $P_{max_{Norm}}$ represent the I_{sc} e P_{Max} normalized to STC; α e γ are the temperature coefficients for the short-circuit current and power, respectively; T_c is the cell temperature; T_{STC} and G_{STC} are the temperature and irradiance at the reference conditions (25 °C and 1,000 W/m^2), and G_{POA} is the incident irradiance at the module plane.

A widely used metric in the scientific community for determining soiling effects is the Soiling Ratio (SRatio), which can be determined using I_{sc} (SRatio$_{Isc}$) or P_{Max} (SRatio$_{PMax}$) data from dirty and clean modules. SRatio$_{PMax}$ has been reported to perform better when measuring nonuniform modules soiling using bypass diodes. SRatio is a dimensionless metric that ranges from 0 to 1; as values approach 1, soiling is considered to have a lesser impact, while values close to 0 indicate a strong soiling effect on PV module performance. For situations using soiling stations (PV modules of the same technology exposed to the same conditions, one clean and the other dirty), SRatio is quantified by Eqs. (2.3) and (2.4). If the use of a soiling station with two identical modules is not possible and a reference PV cell is used, the SRatio can be estimated using Eqs. (2.5) and (2.6) [46].

$$\text{SRatio}_{I_{sc}} = \frac{I_{sc_{Dirty}}}{I_{sc_{Clean}}} \tag{2.3}$$

$$\text{SRatio}_{P_{max}} = \frac{P_{Max_{Dirty}}}{P_{Max_{Clean}}} \tag{2.4}$$

$$\text{SRatio}_{I_{sc}} = \frac{I_{sc_{Sujo}}}{I_{sc_{STC}} \cdot \left[1+\alpha\left(T_{m_{Dirty}} - T_{STC}\right)\right] \cdot \frac{G_{POA}}{G_{STC}}} \tag{2.5}$$

$$\text{SRatio}_{P_{\max}} = \frac{P_{\text{Max}_{\text{Sujo}}}}{P_{\text{Max}_{STC}} \bullet \left[1+\gamma\left(T_{\text{m}_{\text{Dirty}}} - T_{STC}\right)\right] \bullet \frac{G_{POA}}{G_{STC}}} \quad (2.6)$$

Where $I_{sc_{\text{Dirty}}}$ e $I_{sc_{\text{Clean}}}$ are the short-circuit currents of the dirty and clean module, respectively; $P_{\max_{\text{Dirty}}}$ and $P_{\max_{\text{Clean}}}$ are the maximum powers of the dirty and clean module, respectively. Regarding SRatio calculation, even considering correction factors to eliminate the effects of temperature, $\text{SRatio}_{\text{Isc}}$ underestimates soiling loss [47]; hence, clean and dirty modules power should be recorded and compared during the test period [48].

SRatio variation over time is defined as Soiling Rate (SRate), generally expressed in terms of daily losses [44]. SRate values are negative, representing a loss coefficient; the closer they are to 0, the smaller the soiling effect on PV plant performance [49]. To determine SRate, dry periods lasting 14 days or more should be identified, and PV system performance should be estimated using a performance metric. Based on this, the Theil-Sen estimator is used to determine the slope of the performance data (in each dry period) and provide the median of all slopes [45]. This estimator was initially proposed by [50, 51] and is less influenced by anomalies; Theil-Sen regression is determined according to Eq. (2.7).

$$\beta_1 = \text{Median}\left\{b_{i,j} = \frac{t_i - t_j}{x_i - x_j} : x_i \neq x_j, 1 \leq i \leq j \leq n\right\} \quad (2.7)$$

Where (x_i, t_i) and (x_j, t_j) are the bidirectional coordinates of the pairs of points analyzed, and n represents the number of points that make up the sample.

The Performance Ratio (PR), establishing a percentage relationship between the actual PV plant/module generation (E_{CA}) and its theoretical maximum value, is another frequently used metric, ranging from 0 to 1: values close to 1 indicate performance close to the manufacturer's design [49]. PR compares the final yield of the PV plant (Y_F) in relation to the theoretical yield (Y_R) [52], as shown in Eq. (2.8).

$$PR = \frac{Y_F}{Y_R} = \frac{E_{CA}}{P_{\max_{STC}} \bullet \frac{H_{POA}}{G_{STC}}} \quad (2.8)$$

Where E_{CA} is the electrical energy produced (kWh), $P_{\max_{STC}}$ is the power output of the PV plant at STC (kW), H_{POA} total irradiance on the inclined plane of the module (kWh/m^2), and G_{STC} is the reference irradiance at the STC (kW/m^2).

PR depends on T_m, energy dissipation, measurement system, and soiling. Furthermore, PR incorporates effects of shading, electrical losses, module degradation, and solar reflectance. PR calculation, according to Eq. (2.8), is normalized in relation to irradiance and sensitivity to variations of temperature and wind [52].

In such context, the Weather-Corrected Performance Ratio (PR_{Corr}), defined by Eq. (2.9), is proposed by the National Renewable Energy Laboratory (NREL) [53], reflecting the seasonal behavior consistent with PR determination for any PV technology in each site, while the standard PR can vary up to −0.90% for increases of 3 °C in T_m and 3 m/s in wind speed [52].

$$PR_{Corr} = \frac{\sum_i E_{CA_i}}{\sum_i P_{max_{STC}} \cdot \left[1 + \gamma\left(T_{c_{Avg}} - T_{c_i}\right)\right] \cdot \frac{H_{POA_i}}{G_{STC}}} \tag{2.9}$$

Where H_{POA} is the irradiation in the modules plane and $T_{c_{Avg}}$ is the irradiance-weighted average cell temperature, calculated from 1 year of meteorological data, as expressed in Eq. (2.10); T_{ci} is the measured cell temperature estimated according to Eq. (2.11).

$$T_{c_{avg}} = \frac{\sum_i G_{POA_i} \cdot T_{c_i}}{\sum_i G_{POA_i}} \tag{2.10}$$

$$T_{c_i} = T_{m_i} + \frac{G_{POA_i}}{G_{STC}} \cdot \Delta T \tag{2.11}$$

The ΔT parameter is the difference between T_c and the module backsheet and depends on the module construction. Typically, ΔT varies between 2 and 3 °C for flat modules with open-span assembly [54].

References

1. S. Kalogirou, *Solar Energy Engineering: Processes and Systems* (Elsevier, 2009)
2. J.T. Pinho, *Manual de engenharia para Sistemas Fotovoltaicos*,ed by M.A. Galdino, 2nd edn. (CEPEL—CRESESB, 2014)
3. Â. Vian, C.M.V. Tahan, G.J.R. Aguilar, M.R. Gouvea, M.M.F. Gemignani, *Energia Solar: Fundamentos, Tecnologia e Aplicações* (Blucher, 2021)
4. T. Markvart, *Solar Electricity* (Wiley, 2000)
5. H.S. Rauschenbach, *Solar Cell Array Design Handbook* (Springer Netherlands, 1980). https://doi.org/10.1007/978-94-011-7915-7
6. Edumir Physics: How the holes move in a semiconductor? | Edumir Physics (2021, May 21), https://electronicsphysics.com/how-the-holes-move-in-semiconductor/
7. W.A. Beckman, N. Blair, J.A. Duffie, *Solar Engineering of Thermal Processes, Photovoltaics and Wind* (Wiley, 2020)

8. R. Zilles, W.N. Macedo, M.A.B. Galhardo, S.H.F.d. Oliveira, *Sistemas Fotovoltaicos Conectados à Rede Elétrica* (Oficina de Textos, 2012)
9. A.L. Winck, Efeitos da distribuição espectral da radiação solar na caracterização de dispositivos fotovoltaicos de diferentes tecnologias (PhD in Mining, Metallurgical and Materials Engineering), Federal University of Rio Grande do Sul, 2021
10. D.J. Sailor, J. Anand, R.R. King, Photovoltaics in the built environment: A critical review. Energ. Buildings, 111479 (2021). https://doi.org/10.1016/j.enbuild.2021.111479
11. S. Fadhel, C. Delpha, D. Diallo, I. Bahri, A. Migan, M. Trabelsi, M.F. Mimouni, PV shading fault detection and classification based on I-V curve using principal component analysis: Application to isolated PV system. Sol. Energy **179**, 1–10 (2019). https://doi.org/10.1016/j.solener.2018.12.048
12. P.R. Satpathy, R. Sharma, Power and mismatch losses mitigation by a fixed electrical reconfiguration technique for partially shaded photovoltaic arrays. Energy Convers. Manag. **192**, 52–70 (2019). https://doi.org/10.1016/j.enconman.2019.04.039
13. L.d.O. Santos, P.C.M. de Carvalho, C.d.O. Carvalho Filho, Photovoltaic cell operating temperature models: A review of correlations and parameters. IEEE J. Photovoltaics **12**(1), 179–190 (2022). https://doi.org/10.1109/jphotov.2021.3113156
14. L.O. Santos, F.A.A. Souza, C.O.C. Filho, P.C.M. Carvalho, T. AlSkaif, R.I.S. Pereira, Hybrid modeling for photovoltaic module operating temperature estimation. IEEE J. Photovoltaics, 1–9 (2024). https://doi.org/10.1109/jphotov.2024.3372328
15. D.P.N. Nguyen, K. Neyts, J. Lauwaert, Proposed models to improve predicting the operating temperature of different photovoltaic module technologies under various climatic conditions. Appl. Sci. **11**(15), 7064 (2021). https://doi.org/10.3390/app11157064
16. P. Ferrada, D. Olivares, V. del Campo, A. Marzo, F. Araya, E. Cabrera, J. Llanos, J. Correa-Puerta, C. Portillo, D. Román Silva, M. Trigo-Gonzalez, J. Alonso-Montesinos, G. López, J. Polo, F.J. Batlles, E. Fuentealba, Physicochemical characterization of soiling from photovoltaic facilities in arid locations in the Atacama Desert. Sol. Energy **187**, 47–56 (2019). https://doi.org/10.1016/j.solener.2019.05.034
17. G. Raina, S. Sharma, H. Kumar, S. Sinha, Decoding the effects of soiling on bifacial gain from latitude mounted bifacial modules, in *2021 IEEE 2nd International Conference on Smart Technologies for Power, Energy and Control (STPEC)*, (IEEE, 2021). https://doi.org/10.1109/stpec52385.2021.9718682
18. D.S. Braga, Interrelação entre os parâmetros de desempenho e distribuição de sujidade em módulos fotovoltaicos (Master's Degree in Mechanical Engineering), Pontifical Catholic University of Minas Gerais, 2018
19. B. Adothu, S. Kumar, J.J. John, G. Oreski, G. Mathiak, B. Jäckel, V. Alberts, J.B. Jahangir, M.A. Alam, R. Gottschalg, Comprehensive review on performance, reliability, and roadmap of c-Si PV modules in desert climates: A proposal for improved testing standard. Prog. Photovolt. Res. Appl. (2024). https://doi.org/10.1002/pip.3827
20. T. Alkharusi, G. Huang, C.N. Markides, Characterisation of soiling on glass surfaces and their impact on optical and solar photovoltaic performance. Renew. Energy, 119422 (2023). https://doi.org/10.1016/j.renene.2023.119422
21. C. Christian, A. Anderson, C. Baldus-Jeursen, L. Burnham, L. Micheli, D. Parlevliet, E. Pilat, B. Stridh, E. Urrejola, Soiling losses—Impact on the performance of photovoltaic power plants (2022), https://iea-pvps.org/key-topics/soiling-losses-impact-on-the-performance-of-photovoltaic-power-plants/
22. R.R. Rao, M. Mani, P.C. Ramamurthy, An updated review on factors and their inter-linked influences on photovoltaic system performance. Heliyon **4**(9), e00815 (2018). https://doi.org/10.1016/j.heliyon.2018.e00815
23. T. Salamah, A. Ramahi, K. Alamara, A. Juaidi, R. Abdallah, M.A. Abdelkareem, E.-C. Amer, A.G. Olabi, Effect of dust and methods of cleaning on the performance of solar PV module for different climate regions: Comprehensive review. Sci. Total Environ. **827**, 154050 (2022). https://doi.org/10.1016/j.scitotenv.2022.154050

24. G. Picotti, P. Borghesani, M.E. Cholette, G. Manzolini, Soiling of solar collectors—Modelling approaches for airborne dust and its interactions with surfaces. Renew. Sust. Energ. Rev. **81**, 2343–2357 (2018). https://doi.org/10.1016/j.rser.2017.06.043
25. Schwela, D., Morawska, L., & Kotzias, D. (Eds.). (2002). Guidelines for Concentration and Exposure-Response Measurements of Fine and Ultra Fine Particulate Matter for Use in Epidemiological Studies. European Commission Joint Research Centre
26. J.J. John, Characterization of soiling loss on photovoltaic modules, and development of a novel cleaning system (Doctoral thesis), Indian Institute of Technology Bombay 2015
27. B. Figgis, A. Ennaoui, S. Ahzi, Y. Rémond, Review of PV soiling particle mechanics in desert environments. Renew. Sust. Energ. Rev. **76**, 872–881 (2017). https://doi.org/10.1016/j.rser.2017.03.100
28. C.R. Lawrence, J.C. Neff, The contemporary physical and chemical flux of aeolian dust: A synthesis of direct measurements of dust deposition. Chem. Geol. **267**(1–2), 46–63 (2009). https://doi.org/10.1016/j.chemgeo.2009.02.005
29. R.J. Isaifan, D. Johnson, L. Ackermann, B. Figgis, M. Ayoub, Evaluation of the adhesion forces between dust particles and photovoltaic module surfaces. Sol. Energy Mater. Sol. Cells **191**, 413–421 (2019). https://doi.org/10.1016/j.solmat.2018.11.031
30. S. Brambilla, S. Speckart, M.J. Brown, Adhesion and aerodynamic forces for the resuspension of non-spherical particles in outdoor environments. J. Aerosol Sci. **112**, 52–67 (2017). https://doi.org/10.1016/j.jaerosci.2017.07.006
31. K.K. Ilse, B.W. Figgis, V. Naumann, C. Hagendorf, J. Bagdahn, Fundamentals of soiling processes on photovoltaic modules. Renew. Sust. Energ. Rev. **98**, 239–254 (2018). https://doi.org/10.1016/j.rser.2018.09.015
32. T. Sarver, A. Al-Qaraghuli, L.L. Kazmerski, A comprehensive review of the impact of dust on the use of solar energy: History, investigations, results, literature, and mitigation approaches. Renew. Sust. Energ. Rev. **22**, 698–733 (2013b). https://doi.org/10.1016/j.rser.2012.12.065
33. B. Jamil, N. Akhtar, Comparative analysis of diffuse solar radiation models based on sky-clearness index and sunshine period for humid-subtropical climatic region of India: A case study. Renew. Sust. Energ. Rev. **78**, 329–355 (2017). https://doi.org/10.1016/j.rser.2017.04.073
34. S. Nurjanah, T. Dewi, Rusdianasari, Dusting and soiling effect on PV panel performance: Case study open-pit mining in South Sumatra, Indonesia, in *2021 International Conference on Electrical and Information Technology (IEIT)*, (IEEE, 2021). https://doi.org/10.1109/ieit53149.2021.9587351
35. J.G. Bessa, L. Micheli, F. Almonacid, E.F. Fernández, Monitoring photovoltaic soiling: Assessment, challenges, and perspectives of current and potential strategies. IScience **24**(3), 102165 (2021). https://doi.org/10.1016/j.isci.2021.102165
36. H.A. Kazem, M.T. Chaichan, A.H.A. Al-Waeli, K. Sopian, A review of dust accumulation and cleaning methods for solar photovoltaic systems. J. Clean. Prod. **276**, 123187 (2020). https://doi.org/10.1016/j.jclepro.2020.123187
37. K. Styszko, M. Jaszczur, J. Teneta, Q. Hassan, P. Burzyńska, E. Marcinek, N. Łopian, L. Samek, An analysis of the dust deposition on solar photovoltaic modules. Environ. Sci. Pollut. Res. **26**(9), 8393–8401 (2018). https://doi.org/10.1007/s11356-018-1847-z
38. S. L., N. Kumaar, H.P. Waldl, P.K. Das, K. Ramanathan, K. Balaraman, I. Mitra, Impact of soiling on energy yield of solar PV power plant and developing soiling correction factor for solar PV power forecasting. Eur. J. Energy Res. **1**(2), 21–29 (2021). https://doi.org/10.24018/ejenergy.2021.1.2.7
39. M. Jaszczur, Q. Hassan, K. Styszko, J. Teneta, Impact of dust and temperature on energy conversion process in photovoltaic module. Therm. Sci. **23**(Suppl. 4), 1199–1210 (2019). https://doi.org/10.2298/tsci19s4199j
40. B. Figgis, A. Nouviaire, Y. Wubulikasimu, W. Javed, B. Guo, A. Ait-Mokhtar, R. Belarbi, S. Ahzi, Y. Rémond, A. Ennaoui, Investigation of factors affecting condensation on soiled PV modules. Sol. Energy **159**, 488–500 (2018). https://doi.org/10.1016/j.solener.2017.10.089

41. U.C. Laksahani, W.B.I. Udithamala, T.D.K.U. Chathurani, S.L.P. Yasakethu, Design mechanism for solar panel cleaning process, in *2021 10th International Conference on Information and Automation for Sustainability (ICIAFS)*, (IEEE, 2021b). https://doi.org/10.1109/iciafs52090.2021.9606104
42. M. Jaszczur, A. Koshti, W. Nawrot, P. Sędor, An investigation of the dust accumulation on photovoltaic panels. Environ. Sci. Pollut. Res. **27**(2), 2001–2014 (2019). https://doi.org/10.1007/s11356-019-06742-2
43. W.J. Jamil, H. Abdul Rahman, S. Shaari, Z. Salam, Performance degradation of photovoltaic power system: Review on mitigation methods. Renew. Sust. Energ. Rev. **67**, 876–891 (2017). https://doi.org/10.1016/j.rser.2016.09.072
44. R. Conceição, J. González-Aguilar, A.A. Merrouni, M. Romero, Soiling effect in solar energy conversion systems: A review. Renew. Sust. Energ. Rev. **162**, 112434 (2022). https://doi.org/10.1016/j.rser.2022.112434
45. S.C.S. Costa, Estudo abrangente do efeito da sujidade no desempenho de módulos e sistemas fotovoltaicos (PhD in Mechanical Engineering), Pontifical Catholic University of Minas, 2018
46. M. Gostein, T. Duster, C. Thuman, Accurately measuring PV soiling losses with soiling station employing module power measurements, in *2015 IEEE 42nd Photovoltaic Specialists Conference (PVSC)*, (IEEE, 2015). https://doi.org/10.1109/pvsc.2015.7355993
47. M. Gostein, B. Littmann, J.R. Caron, L. Dunn, Comparing PV power plant soiling measurements extracted from PV module irradiance and power measurements, in *2013 IEEE 39th photovoltaic specialists conference (PVSC)*, (IEEE, 2013b). https://doi.org/10.1109/pvsc.2013.6745094
48. H. Yazdani, M. Yaghoubi, Dust deposition effect on photovoltaic modules performance and optimization of cleaning period: A combined experimental–numerical study. Sustain Energy Technol Assess **51**, 101946 (2022). https://doi.org/10.1016/j.seta.2021.101946
49. D.N. Araújo, Investigação experimental dos efeitos da sujidade no desempenho de plantas fotovoltaicas instaladas no campus do Pici da UFC (Master's in Electrical Engineering não publicada), Federal University of Ceará, 2019
50. H. Theil, A rank-invariant method of linear and polynomial regression analysis, in *Advanced Studies in Theoretical and Applied Econometrics*, (Springer Netherlands, 1992), pp. 345–381. https://doi.org/10.1007/978-94-011-2546-8_20
51. P.K. Sen, Estimates of the regression coefficient based on kendall's tau. J. Am. Stat. Assoc. **63**(324), 1379–1389 (1968). https://doi.org/10.1080/01621459.1968.10480934
52. E. Urrejola, J. Antonanzas, P. Ayala, M. Salgado, G. Ramírez-Sagner, C. Cortés, A. Pino, R. Escobar, Effect of soiling and sunlight exposure on the performance ratio of photovoltaic technologies in Santiago, Chile. Energy Convers. Manag. **114**, 338–347 (2016). https://doi.org/10.1016/j.enconman.2016.02.016
53. T. Dierauf, A. Growitz, S. Kurtz, J.L.B. Cruz, E. Riley, C. Hansen, *Weather-Corrected Performance Ratio* (Office of Scientific and Technical Information (OSTI), 2013). https://doi.org/10.2172/1078057
54. J.A. Kratochvil, W.E. Boyson, D.L. King, *Photovoltaic Array Performance Model* (Office of Scientific and Technical Information (OSTI), 2004). https://doi.org/10.2172/919131

Chapter 3
State–of-the-Art Perspectives

3.1 Planning and Searches

A Systematic Literature Review (SLR) is a methodology that helps guide the search for information by establishing search parameters, inclusion/exclusion criteria, and how the information will be processed [1]. This SLR uses the StArt (State of the Art Through Systematic Review) tool [2]; the process is divided into six steps:

1. **Planning**: defining objective and research questions, selecting repositories, establishing selection criteria, and choosing keywords.
2. **Search and Identification**: applying the search term to the selected repositories, manually inserting studies indicated by the study group with potential for selection, and excluding duplicate studies.
3. **Primary Selection**: reading the title and abstract of all identified Primary Studies (PS) and applying the inclusion and exclusion criteria.
4. **Secondary Selection**: superficial reading of the methodology, results, and conclusions of the studies selected in the previous step; reapplication of inclusion and exclusion criteria.
5. **Extraction**: thorough reading of all selected PS to answer the research questions.
6. **Summarization**: qualitative and quantitative analysis of the selected studies.

Regarding search strategies, the Engineering Village, IEEE Xplore, Scopus, and Web of Science repositories were used due to extensive collections. The studies analyzed were limited to the period 2016–2024.

Keywords are the most used strategy searching for PS [1]. Furthermore selection criteria can be established through the composition of search expressions constructed by combining keywords with Boolean operators. To narrow the searches filters are applied in each repository. The keywords and their synonyms used to

J. J. Silva de Souza, P. C. Marques de Carvalho, *Soiling Effects on Photovoltaic Systems*, SpringerBriefs in Applied Sciences and Technology,
https://doi.org/10.1007/978-3-032-22442-2_3

Table 3.1 Keywords used in searches

Keywords	Synonyms
Photovoltaic	*Solar energy* *PV*
Soiling	*Dust* *Dirt*
Effect	*Impact* *Influence*
Characteristic	*Property*
Analyze	*Investigation* *Study* *Assessment* *Evaluation* *Research*

construct the search strings are shown in Table 3.1. The terms were chosen after prior analysis of literature reviews on soiling such as the studies by [3–10].

One of the fundamental SLR characteristics is the prior definition of the selection criteria (inclusion and exclusion) for PS; hence, to establish in advance whether the inclusion of a PS is conditional on the application of one or more inclusion criteria is essential [11]. For this SLR, four inclusion criteria (I1)–(I4) are established and should be applied together for the admission of a PS:

- **(I1)**: Studies available in full *AND*.
- **(I2)**: Experimental studies under outdoor conditions that describe the physico-chemical characteristics of soiling and the effects of its natural accumulation on PV performance *OR*.
- **(I3)**: Experimental studies under outdoor conditions describe the effects of natural soiling accumulation on PV performance, even without mentioning the physicochemical characteristics of soiling *OR*.
- **(I4)**: Experimental studies that describe the characteristics of the physical-chemical composition of soiling on PV systems or glass samples, even without mentioning PV performance.

Exclusion criteria define the characteristics used to identify PS that are irrelevant to an SLR, such as failure to meet the inclusion criteria. In this study, six exclusion criteria (E1)–(E6) are adopted, which, when applied in isolation, are sufficient to disqualify a PS from this SLR:

- **(E1)**: Studies that describe soiling effects on PV performance only under artificial deposition/environmental/soiling conditions *OR*.
- **(E2)**: Studies that analyze impacts only through simulations and/or mathematical models *OR*.
- **(E3)**: Studies that investigate soiling effects only on Concentrated Solar Power (CSP) systems *OR*.

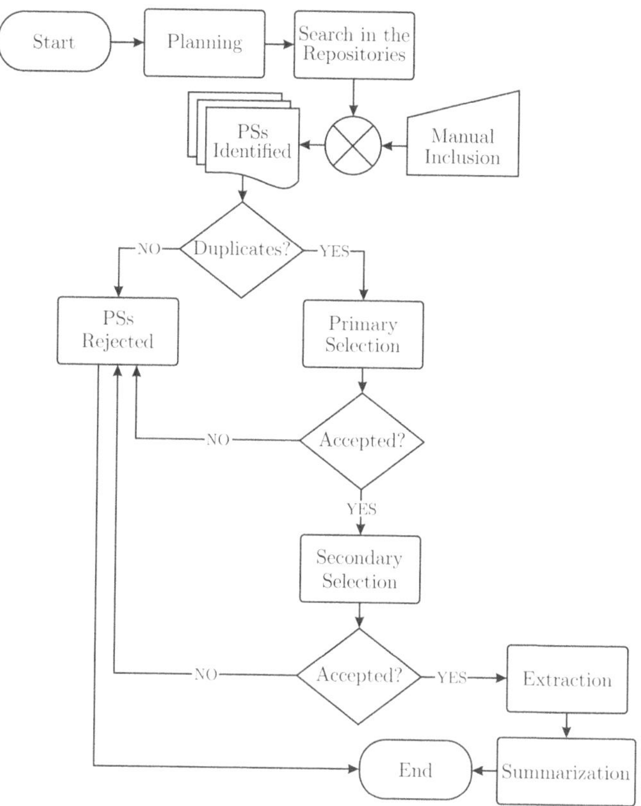

Fig. 3.1 Flowchart for RSL execution

- **(E4)**: Studies published only as abstracts or posters *OR*.
- **(E5)**: Literature review papers *OR*.
- **(E6)**: Papers that do not meet the inclusion criteria.

The flowchart of the steps adopted in this SRL is shown in Fig. 3.1.

3.2 PS Analysis and Discussions

Aiming to synthesize and provide a better understanding of research trends, PS analysis is presented in the following thematic areas: (1) PV technologies used, (2) methodologies adopted in the studies, (3) cleaning methods, (4) soiling effects, and (5) physicochemical characteristics.

3.2.1 PV Technologies

Poly (p-Si) and monocrystalline silicon (m-Si) are the most used modules in soiling studies, representing approximately 65.40% of the analyzed PV modules. More than 95% of the PV modules available on the global market use crystalline silicon technology [12]; therefore, understanding soiling impacts on such modules in different sites can contribute to the development and implementation of soiling mitigation strategies, in addition to enabling module construction optimization and formulation of recommendations for soiling management. Other technologies, such as Cadmium Telluride (CdTe) and Copper Indium Gallium Diselenide (CIGS) are also used; studies on organic PV cells [13, 14] and mono and bifacial heterojunction cells [15] are also identified. For a better understanding of soiling effects, the use of different PV technologies should be considered, as in [16–20], since the results from a given technology should not be extrapolated, due to differences in spectral responses [21, 22]. Another aspect that should be considered is the use of bifacial modules since the deposition process and the need for mitigation occur differently in relation to monofacial modules [16, 23–25]. In addition, the installation mode (portrait or pass-through) [26], geometry [18, 19], and module material are factors that influence soiling effects.

3.2.2 PS Methodologies

Approximately 57% of the studies consist of comparing PV modules or plants performance subjected to manual and natural cleaning [27–30]. Approximately 22% of the PS investigates PV modules or plants subjected only to natural cleaning, evaluating soiling effects through metrics such as SRatio and PR/PR_{Corr} during the observation period and using climatic/environmental/operational data to determine seasonal behavior [22, 31–33]. Furthermore, studies have been conducted using anti-fouling coatings [34–36], automatic cleaning systems [37], and robotic systems [38–40]. Regarding anti-fouling coatings use, additional studies on the effectiveness of this strategy are necessary, especially in desert conditions, due to factors such as water unavailability and difficult access to PV modules; aging caused by exposure to desert conditions, such as high irradiance and temperature, should also be investigated when analyzing the applicability of this soiling mitigation methodology.

3.2.3 Cleaning Methods

Water is the commonly used option for manual cleaning; however, in desert areas, access to this resource is a limiting factor. Therefore, the use of dry/vibration cleaning methods [41–43] anti-soil coatings, and water recycling systems [44] may be viable solutions. Depending on the soiling characteristics, these methods may have limited effectiveness or even enhance the effects, as when using anti-soil coatings in South Africa [34, 35]. In general, the choice of the best mitigation method depends on soiling characteristics, resources availability, implementation costs, and cleaning frequency. The cleaning interval is influenced by PV modules installation angle [24]; according to the literature, leaving PV modules dirty can shift the ideal installation angle [45, 46] Furthermore, a factor to be considered when defining the soiling mitigation method and the frequency of interventions is the technical and financial feasibility, aiming to reduce PV plants financial risk due to the demand for specialized labor for cleaning interventions.

Another important aspect for soiling mitigation is the natural cleaning; during periods of frequent rainfall, soiling losses tend to be lower, reducing the need for cleaning interventions [47], except for uneven deposition, especially in framed modules, to avoid hot spots [48]. Therefore, precipitation is an important self-cleaning mechanism for PV systems [49–51]; its effectiveness depends on the frequency and volume of precipitation [47, 52–54], the previous dry period [55], extreme events [56], and the tilt angle [24]. Low-intensity precipitation can exacerbate soiling effects [57]; due to the physical and chemical nature of soiling particles, frequent precipitation may not be sufficient to keep PV modules clean [58]. Another effective natural agent in the self-cleaning process is the wind [37, 54, 59], especially when combined with low humidity [60]; wind contributes to the resuspension of soiling deposited on the modules [37, 54]. Regarding humidity, dew formation may lead to greater accumulation of soiling on modules surface [61, 62], causing non-uniform deposition [63] or resulting in a self-cleaning effect [64]; dew also influences the choice of the best time to perform cleaning interventions [65].

3.2.4 Soiling Effects

Irregularly deposited soiling causes PV cell temperature irregularities (hot spots) [48, 66] however, uniformly soiled modules may have lower temperatures than clean modules under the same environmental conditions [18, 67–70] since soiling reduces solar irradiation transmission to PV cells [17, 42]. Clean modules have higher thermal losses due to lower solar irradiation reflectance [18]; however, the T_m reduction effect of the dirty module may be limited by soiling properties [71], being insignificant in some locations [72]. Regarding the installation angle, an inverse relationship with soiling is identified: the smaller the angle, the greater the soiling deposition [44, 49, 73, 74]; furthermore, the deposition pattern [63] and sludge formation are related to the inclination angle. Considering the influence of climatic, environmental, operational, and anthropogenic factors on the deposition process, studies on soiling impacts should be conducted outdoors, with exposure to the solar spectrum and naturally deposited soiling, for a period that reflects local seasonal changes. While studies conducted under laboratory conditions provide relevant information, such investigations cannot adequately represent all atmospheric conditions and the complex interconnections between different variables and soiling.

Soiling effects can be assessed by directly monitoring PV modules electrical parameters subjected to natural cleaning or by comparing data from dirty (naturally cleaned only) and clean (subjected to active or passive cleaning methods) modules; the main parameters are power, electricity production, I_{sc}, and V_{oc}. Another methodology used is the adoption of evaluation metrics, which, in addition to using electrical data, can include information regarding climatic, environmental, and operational variables to standardize results for STC conditions; the main metrics are PR/PR_{Corr}, efficiency, SRatio, and SRate; FF, CF, and productivity (Yield) are also used. Assessment of soiling effects on PV electrical performance is a complex process, depending on several factors. Therefore, additional care should be taken when determining evaluation metrics to avoid underestimating or overestimating the results. Using I_{sc} to determine the SRatio can lead to errors if soiling is not uniformly distributed across the module surface [26]; power data is more recommended for Si modules due to the action of bypass diodes [21]. However, in the case of uniform deposition on p-Si modules, SRatio values based on power and I_{sc} data are similar [21].

Despite the difficulty in generalizing soiling impacts on PV electrical parameters, V_{oc} is generally the least impacted variable, presenting little variation even with increasing exposure time; according to [75, 76], soiling effect on V_{oc} can be neglected. On the other hand, I_{sc} and power are generally the most impacted parameters; as soiling increases, a greater reduction of these parameters is observed since soiling blocks part of the solar irradiation [77, 78]. However, in p-Si modules, non-uniform deposition has a greater impact, as the internal diodes act to isolate the cells compromised by soiling (typically concentrated at the bottom due to the modules inclination) [79]. Furthermore, soiling impacts on PV performance depend on its properties and prevailing climatic conditions [73, 80]. Soiling accumulation affects light transmission in modules, causing irradiance attenuation at different wavelengths. The greatest losses are related to low irradiance levels [81], becoming greater for p-Si technology, compared to m-Si, for irradiance levels below 500 W/m^2 [82]. Nevertheless, the greatest power reductions are associated with small installation angles [24, 83, 84]. Additionally, PV modules power output tends to gradually decrease with increasing exposure time and weather conditions [85]. However, the exposure time provides no sufficient information to assess the reduction of the incident irradiation due to soiling deposition [73].

3.2.5 *Physicochemical Characteristics*

The concentration and type of elements present in the soiling composition depend on the climatic and geological characteristics of the PV plant installation site [86], the occurrence of unusual events [87], and the exposure period [88]. Si, Ca, Al, Fe, O, and Mg are commonly present in the soiling; nevertheless, significant variability is observed, with quartz, calcite, hematite, dolomite, and calcium oxide being the most frequently reported. Similarities are observed between soiling samples collected from PV modules surface and from installation site soil, with differences in the intensity of the elements' occurrence [86, 89]. However, particles of remote origin play a significant role in PV performance, as soiling does not originate solely from local events [90].

Regarding particle size, a range of 0.01 μm [91] to 500 μm [46] is observed; particles smaller than 10 μm are difficult to remove [92]. Small particles are associated with long exposure periods, while larger particles are attributed to short exposure periods and the occurrence of sandstorms [87, 88]. The morphology changes with increasing exposure time [87, 88]. Particle agglomeration is associated with long exposure times, as particles tend to be small [87]. Needle-shaped particles are identified as the main responsible for dirt cementation on modules surface [93] and are predominantly formed during dew events due to the dissolution-precipitation process [94].

3.2.6 Further Discussion

Considering the developed SLR, soiling negatively affects PV plants and exhibits distinct behavior in different regions. Therefore, accurately modeling soiling losses, considering particles with different physicochemical-biological characteristics and distinct PV technologies, is a difficult task. Furthermore, most of the analyzed studies focus on impacts of soiling on PV plants electrical parameters; the physicochemical analysis of soiling, considering periods with different climatic regimes, is limited. Although different studies have been conducted globally, investigations into the impacts of soiling are challenging and still limited, as they depend on specific local climatic conditions. In addition, the impacts of anthropogenic activities in the vicinity of PV plants should be considered since they represent an important source of soiling particles and are commonly neglected during the implementation phase of PV systems. In this sense, assessments of different mitigation processes should be systematically analyzed, aiming to maximize PV production, especially in large urban environments, since the literature indicates significant gaps in the investigation of the pollution effect on PV performance.

In the short term, the adoption of cleaning schedules can reduce the effects of contaminant deposition on PV generation, especially in installations on rooftops in highly polluted urban areas. In the long term, pollution mitigation and the recovery of desertified areas can contribute to the reduction of atmospheric aerosols and, consequently, the rate of dirt deposition [95]. Furthermore, dirt tends to continue to be a problem of global interest due to climate change, the increase in global temperature, and the occurrence of prolonged periods of drought. Thus, studies focused on modeling, adaptation, and mitigation strategies for soiling are essential to maximize PV plants generation and financial and economic gains [96].

References

1. S. Ribeiro, E. Schmitz, A. Alencar, M. Silva, Literature review on the theory of constraints applied in the software development process. IEEE Lat. Am. Trans. **16**(11), 2747–2756 (2018). https://doi.org/10.1109/tla.2018.8795116
2. E. Hernandes, A. Zamboni, S. Fabbri, A. Di Thommazo, Using GQM and TAM to evaluate StArt—A tool that supports systematic review. CLEI Electr. J. **15**(1) (2012). https://doi.org/10.19153/cleiej.15.1.2
3. Y.N. Chanchangi, A. Ghosh, S. Sundaram, T.K. Mallick, Dust and PV performance in Nigeria: A review. Renew. Sust. Energ. Rev. **121**, 109704 (2020). https://doi.org/10.1016/j.rser.2020.109704
4. Z.A. Darwish, H.A. Kazem, K. Sopian, M.A. Al-Goul, H. Alawadhi, Effect of dust pollutant type on photovoltaic performance. Renew. Sust. Energ. Rev. **41**, 735–744 (2015). https://doi.org/10.1016/j.rser.2014.08.068
5. M.M. Fouad, L.A. Shihata, E.I. Morgan, An integrated review of factors influencing the perfomance of photovoltaic panels. Renew. Sust. Energ. Rev. **80**, 1499–1511 (2017). https://doi.org/10.1016/j.rser.2017.05.141
6. A. Gholami, M. Ameri, M. Zandi, R.G. Ghoachani, S. Eslami, S. Pierfederici, Photovoltaic potential assessment and dust impacts on photovoltaic systems in Iran: Review paper. IEEE J. Photovoltaics **10**(3), 824–837 (2020). https://doi.org/10.1109/jphotov.2020.2978851
7. M.R. Maghami, H. Hizam, C. Gomes, M.A. Radzi, M.I. Rezadad, S. Hajighorbani, Power loss due to soiling on solar panel: A review. Renew. Sust. Energ. Rev. **59**, 1307–1316 (2016). https://doi.org/10.1016/j.rser.2016.01.044
8. M. Mani, R. Pillai, Impact of dust on solar photovoltaic (PV) performance: Research status, challenges and recommendations. Renew. Sust. Energ. Rev. **14**(9), 3124–3131 (2010). https://doi.org/10.1016/j.rser.2010.07.065
9. M. Mussard, M. Amara, Performance of solar photovoltaic modules under arid climatic conditions: A review. Sol. Energy **174**, 409–421 (2018). https://doi.org/10.1016/j.solener.2018.08.071
10. M. Santhakumari, N. Sagar, A review of the environmental factors degrading the performance of silicon wafer-based photovoltaic modules: Failure detection methods and essential mitigation techniques. Renew. Sust. Energ. Rev. **110**, 83–100 (2019). https://doi.org/10.1016/j.rser.2019.04.024
11. K.R.F. Scannavino, E.Y. Nakagawa, S.C.P.F. Fabbri, F.C. Ferrari, *Revisão Sistemática da Literatura em Engenharia de Software: teoria e prática* (Elsevier, Rio de Janeiro, 2017)
12. International Energy Agency—IEA: Solar PV, Paris (2022), https://www.iea.org/energy-system/renewables/solar-pv. Accessed 19 June 2025
13. W. Luo, A.M. Khaing, C.D. Rodriguez-Gallegos, S.W. Leow, T. Reindl, M. Pravettoni, Long-term outdoor study of an organic photovoltaic module for building integration. Prog. Photovolt. Res. Appl. (2024). https://doi.org/10.1002/pip.3791
14. G. Touhami, L. Slimane, A. Fatima Zohra, B. Bouchra, Evaluation experimental of the impact of Saharan climate conditions on the infinity organic photovoltaic module performance. Aust. J. Electr. Electron. Eng., 1–11 (2022). https://doi.org/10.1080/1448837x.2022.2108585
15. A.A. Abdallah, M. Kivambe, B. Aïssa, B.W. Figgis, Performance of monofacial and bifacial silicon heterojunction modules under desert conditions and the impact of PV soiling. Sustainability **15**(10), 8436 (2023). https://doi.org/10.3390/su15108436
16. A. Baras, R.K. Jones, A. Alqahtani, M. Alodan, K. Abdullah, Measured soiling loss and its economic impact for PV plants in central Saudi Arabia, in *2016 Saudi Arabia Smart Grid (SASG)*, (IEEE, 2016). https://doi.org/10.1109/sasg.2016.7849657
17. S. Kaundilya, O.S. Sastry, B. Bora, S. Rai, M. Bangar, Renu, R. Singh, A. Kumar, K. Yadav, M. Kumar, R. Arun Prasath, Soiling effect on crystalline and thin-film technology PV modules for composite climate zone of India. Mater. Today Proc. **5**(11), 23275–23280 (2018). https://doi.org/10.1016/j.matpr.2018.11.060

18. M. Adouane, A. Al-Qattan, B. Alabdulrazzaq, A. Fakhraldeen, Comparative performance evaluation of different photovoltaic modules technologies under Kuwait harsh climatic conditions. Energy Rep. **6**, 2689–2696 (2020). https://doi.org/10.1016/j.egyr.2020.09.034
19. B. Alabdulrazzaq, M. Adouane, A. Al-Qattan, On the effect of PV geometry on soiling: Exploring use-cases for cylindrical PV modules as a soiling loss mitigation method, in *2020 IEEE 47th Photovoltaic Specialists Conference (PVSC)*, (IEEE, 2020). https://doi.org/10.1109/pvsc45281.2020.9300771
20. R. Shadid, Y. Khawaja, A. Bani-Abdullah, M. Akho-Zahieh, A. Allahham, Investigation of weather conditions on the output power of various photovoltaic systems. Renew. Energy, 119202 (2023). https://doi.org/10.1016/j.renene.2023.119202
21. S.C.S. Costa, A.S.A.C. Diniz, V.A.C. Santana, Muller, Matthew, L. Micheli, L. Kazmerski, *Avaliação da sujidade em módulos fotovoltaicos em Minas Gerais, Brasil* (Anais Congresso Brasileiro De Energia Solar—CBENS, 2018). https://doi.org/10.59627/cbens.2018.191
22. S.C.S. Costa, A.S.A.C. Diniz, D.S. Braga, V.A.C. Santana, T. Duarte, L.L. Kazmerski, Effects of soiling of photovoltaic modules and systems in Brazil's climate zones, in *ISES Solar World Congress 2019/IEA SHC International Conference on Solar Heating and Cooling for Buildings and Industry 2019*, (International Solar Energy Society, 2019). https://doi.org/10.18086/swc.2019.16.07
23. P. Ayala, C. Munoz, N. Osorio, C. Hernandez, F. Zurita, V. Gutierrez, G. Ramirez, F. Mancilla, P. Valdivia, F. Cuevas, P. Ferrada, Bifacial technology performance compared with three commercial monofacial PV technologies under outdoor high irradiance conditions at the Atacama desert, in *2018 IEEE 7th World Conference on Photovoltaic Energy Conversion (WCPEC) (A Joint Conference of 45th IEEE PVSC, 28th PVSEC & 34th EU PVSEC)*, (IEEE, 2018). https://doi.org/10.1109/pvsc.2018.8547345
24. A. Ullah, A. Amin, T. Haider, M. Saleem, N.Z. Butt, Investigation of soiling effects, dust chemistry and optimum cleaning schedule for PV modules in Lahore, Pakistan. Renew. Energy **150**, 456–468 (2020). https://doi.org/10.1016/j.renene.2019.12.090
25. G. Raina, S. Sinha, Experimental investigations of front and rear side soiling on bifacial PV module under different installations and environmental conditions. Energy Sustain. Dev. **72**, 301–313 (2023). https://doi.org/10.1016/j.esd.2023.01.001
26. S. Kagan, E. Giosa, R. Flottemesch, R. Andrews, J. Rand, M. Reed, M. Gostein, B. Stueve, Impact of non-uniform soiling on PV system performance and soiling measurement, in *2018 IEEE 7th World Conference on Photovoltaic Energy Conversion (WCPEC) (A Joint Conference of 45th IEEE PVSC, 28th PVSEC & 34th EU PVSEC)*, (IEEE, 2018). https://doi.org/10.1109/pvsc.2018.8547728
27. M. Jaszczur, Q. Hassan, J. Teneta, K. Styszko, W. Nawrot, R. Hanus, Study of dust deposition and temperature impact on solar photovoltaic module. MATEC Web Conf. **240**, 04005 (2018). https://doi.org/10.1051/matecconf/201824004005
28. S. Bhaduri, A. Kottantharayil, Mitigation of soiling by vertical mounting of bifacial modules. IEEE J. Photovoltaics **9**(1), 240–244 (2019). https://doi.org/10.1109/jphotov.2018.2872555
29. W. Jo, N. Ham, J. Kim, J. Kim, The cleaning effect of photovoltaic modules according to precipitation in the operation stage of a large-scale solar power plant. Energies **16**(17), 6180 (2023). https://doi.org/10.3390/en16176180
30. J.J.S. de Souza, P.C.M. de Carvalho, *Estimativa da taxa de sujidade da planta fotovoltaica do IFCE—campus Cedro: um estudo de caso* (Anais Congresso Brasileiro De Energia Solar—CBENS, 2024). https://doi.org/10.59627/cbens.2024.2341
31. A.K. Tripathi, M. Aruna, C.S.N. Murthy, Performance evaluation of PV panel under dusty condition. Int. J. Renewable Energy Dev. **6**(3), 225 (2017). https://doi.org/10.14710/ijred.6.3.225-233
32. F. Touati, N.A. Chowdhury, K. Benhmed, A.J.R. San Pedro Gonzales, M.A. Al-Hitmi, M. Benammar, A. Gastli, L. Ben-Brahim, Long-term performance analysis and power prediction of PV technology in the State of Qatar. Renew. Energy **113**, 952–965 (2017). https://doi.org/10.1016/j.renene.2017.06.078

33. D.N. Norberto Araujo, S.C.S. Costa, I.M. Dupont, P.C.M. CARVALHO, *Impactos da sujidade e efeitos da limpeza no desempenho de módulos fotovoltaicos* (Anais Congresso Brasileiro De Energia Solar—CBENS, 2020). https://doi.org/10.59627/cbens.2020.862
34. A.A. Plessis, J.M. Strauss, A.J. Rix, Application of dust mitigation strategies to single-axis-tracking photovoltaic modules in the semi-arid areas of South Africa, in *2019 International Conference on Clean Electrical Power (ICCEP)*, (IEEE, 2019). https://doi.org/10.1109/iccep.2019.8890145
35. A.A. Plessis, J.M. Strauss, A.J. Rix, Application of dust mitigation strategies to single-axis-tracking photovoltaic modules in the semi-arid areas of South Africa. IET Renewable Power Gen. **14**(15), 2781–2790 (2020). https://doi.org/10.1049/iet-rpg.2019.1424
36. H. Al Bakri, W. Abu Elhaija, A. Al Zyoud, Solar photovoltaic panels performance improvement using active self-cleaning nanotechnology of SurfaShield G. Energy **223**, 119908 (2021). https://doi.org/10.1016/j.energy.2021.119908
37. S. Caroline Silva Costa, A. Sonia Alves Cardoso Diniz, V. Augusto Camatta Santana, L. Kazmerski, *Determinação das taxas de sujidade para módulos fotovoltaicos de filme fino e silício cristalino instalados em diferentes zonas climáticas brasileiras* (Anais Congresso Brasileiro De Energia Solar—CBENS, 2020). https://doi.org/10.59627/cbens.2020.821
38. S. Tiwari, P. Rani, R.N. Patel, Examining the economic viability of a solar panel dust cleaning, in *2019 IEEE International Conference on Electrical, Computer and Communication Technologies (ICECCT)*, (IEEE, 2019). https://doi.org/10.1109/icecct.2019.8869011
39. M.K. Ghodki, A. Swarup, Y. Pal, A new IR and sprinkler based embedded controller directed robotic arm for automatic cleaning of solar panel. J. Eng., Des. Technol. **18**(4), 905–921 (2019). https://doi.org/10.1108/jedt-10-2019-0253
40. M. Hammoud, B. Shokr, A. Assi, J. Hallal, P. Khoury, Effect of dust cleaning on the enhancement of the power generation of a coastal PV-power plant at Zahrani Lebanon. Sol. Energy **184**, 195–201 (2019b). https://doi.org/10.1016/j.solener.2019.04.005
41. D. Dahlioui, B. Laarabi, A. Barhdadi, Investigation of soiling impact on PV modules performance in semi-arid and hyper-arid climates in Morocco. Energy Sustain. Dev. **51**, 32–39 (2019). https://doi.org/10.1016/j.esd.2019.05.001
42. M. Al-Housani, Y. Bicer, M. Koç, Assessment of various dry photovoltaic cleaning techniques and frequencies on the power output of cdte-type modules in dusty environments. Sustainability **11**(10), 2850 (2019). https://doi.org/10.3390/su11102850
43. R. Shenouda, M.S. Abd-Elhady, A.R. Ali, H. Kandil, Influence of vibration time on dust accumulation on PV panels that operate light posts. Energy Rep. **8**, 309–318 (2022). https://doi.org/10.1016/j.egyr.2022.10.151
44. R. Majeed, A. Waqas, H. Sami, M. Ali, N. Shahzad, Experimental investigation of soiling losses and a novel cost-effective cleaning system for PV modules. Sol. Energy **201**, 298–306 (2020). https://doi.org/10.1016/j.solener.2020.03.014
45. E. Abdeen, M. Orabi, E.-S. Hasaneen, Optimum tilt angle for photovoltaic system in desert environment. Sol. Energy **155**, 267–280 (2017). https://doi.org/10.1016/j.solener.2017.06.031
46. M. Abdolzadeh, R. Nikkhah, Experimental study of dust deposition settled over tilted PV modules fixed in different directions in the southeast of Iran. Environ. Sci. Pollut. Res. **26**(30), 31478–31490 (2019). https://doi.org/10.1007/s11356-019-06246-z
47. M.A.M.L.d. Jesus, G. Timò, C. Agustín-Sáenz, I. Braceras, M. Cornelli, A.d.M. Ferreira, Anti-soiling coatings for solar cell cover glass: Climate and surface properties influence. Sol. Energy Mater. Sol. Cells **185**, 517–523 (2018). https://doi.org/10.1016/j.solmat.2018.05.036
48. S.C.S. Costa, L.L. Kazmerski, A.S.A.C. Diniz, Impact of soiling on Si and CdTe PV modules: Case study in different Brazil climate zones. Energy Conv. Manage.: X **10**, 100084 (2021). https://doi.org/10.1016/j.ecmx.2021.100084
49. J. Tanesab, D. Parlevliet, J. Whale, T. Urmee, Seasonal effect of dust on the degradation of PV modules performance deployed in different climate areas. Renew. Energy **111**, 105–115 (2017). https://doi.org/10.1016/j.renene.2017.03.091

50. E. Urrejola, J. Antonanzas, P. Ayala, M. Salgado, G. Ramírez-Sagner, C. Cortés, A. Pino, R. Escobar, Effect of soiling and sunlight exposure on the performance ratio of photovoltaic technologies in Santiago, Chile. Energy Convers. Manag. **114**, 338–347 (2016). https://doi.org/10.1016/j.enconman.2016.02.016
51. E.A. Tonolo, J.D. Mariano, J.U. Junior, *Análise do efeito do acúmulo de sujeira nos sistemas fotovoltaicos da UTFPR—câmpus Curitiba* (Anais Congresso Brasileiro De Energia Solar—CBENS, 2018). https://doi.org/10.59627/cbens.2018.183
52. D.H. Daher, L. Gaillard, M. Amara, C. Ménézo, Impact of tropical desert maritime climate on the performance of a PV grid-connected power plant. Renew. Energy **125**, 729–737 (2018). https://doi.org/10.1016/j.renene.2018.03.013
53. K. Chiteka, R. Arora, S.N. Sridhara, C.C. Enweremadu, A novel approach to solar PV cleaning frequency optimization for soiling mitigation. Sci. Afr. **8**, e00459 (2020). https://doi.org/10.1016/j.sciaf.2020.e00459
54. W. Javed, B. Guo, B. Figgis, L. Martin Pomares, B. Aïssa, Multi-year field assessment of seasonal variability of photovoltaic soiling and environmental factors in a desert environment. Sol. Energy **211**, 1392–1402 (2020). https://doi.org/10.1016/j.solener.2020.10.076
55. W. Javed, B. Guo, Y. Wubulikasimu, B.W. Figgis, Photovoltaic performance degradation due to soiling and characterization of the accumulated dust, in *2016 IEEE International Conference on Power and Renewable Energy (ICPRE)*, (IEEE, 2016). https://doi.org/10.1109/icpre.2016.7871142
56. H. Yazdani, M. Yaghoubi, Techno-economic study of photovoltaic systems performance in Shiraz, Iran. Renewable Energy **172**, 251–262 (2021). https://doi.org/10.1016/j.renene.2021.03.012
57. E.R.A. Larico, J.M.R. Cutipa, L.V. Callata, Effect of dust and rain on the performance of a photovoltaic system at more than 3800 m altitude, in *2020 IEEE Engineering International Research Conference (EIRCON)*, (IEEE, 2020). https://doi.org/10.1109/eircon51178.2020.9254058
58. J.G. Bessa, M. Valerino, M. Muller, M. Bergin, L. Micheli, F. Almonacid, E.F. Fernández, An investigation on the pollen-induced soiling losses in utility-scale PV plants. IEEE J. Photovoltaics, 1–7 (2023). https://doi.org/10.1109/jphotov.2023.3326560
59. H.R. Alamri, H. Rezk, H. Abd-Elbary, H.A. Ziedan, A. Elnozahy, Experimental investigation to improve the energy efficiency of solar PV panels using hydrophobic sio2 nanomaterial. Coatings **10**(5), 503 (2020). https://doi.org/10.3390/coatings10050503
60. N.K. Kasim, N.M. Obaid, H.G. Abood, R.A. Mahdi, A.M. Humada, Experimental study for the effect of dust cleaning on the performance of grid-tied photovoltaic solar systems. Int. J. Electr. Comput. Eng. (IJECE) **11**(1), 74 (2021b). https://doi.org/10.11591/ijece.v11i1.pp74-83
61. S.C.S. Costa, A.M.F.V. Abreu, M.M. Viana, P.P. Brito, M.V. de Assis, C.B. Maia, A.S.A.C. Diniz, L.L. Kazmerski, *Caracterização físico-química da sujidade depositada sobre módulos fotovoltaicos instalados em zonas climáticas de Minas Gerais* (Anais Congresso Brasileiro De Energia Solar—CBENS, 2016), pp. 1–8. https://doi.org/10.59627/cbens.2016.1461
62. M. Al-Housani, Y. Bicer, M. Koç, Experimental investigations on PV cleaning of large-scale solar power plants in desert climates: Comparison of cleaning techniques for drone retrofitting. Energy Convers. Manag. **185**, 800–815 (2019). https://doi.org/10.1016/j.enconman.2019.01.058
63. A. Azouzoute, C. Hajjaj, H. Zitouni, M. El Ydrissi, O. Mertah, M. Garoum, A. Ghennioui, Modeling and experimental investigation of dust effect on glass cover PV module with fixed and tracking system under semi-arid climate. Sol. Energy Mater. Sol. Cells **230**, 111219 (2021). https://doi.org/10.1016/j.solmat.2021.111219
64. B.R. Paudyal, S.R. Shakya, Dust accumulation effects on efficiency of solar PV modules for off grid purpose: A case study of Kathmandu. Sol. Energy **135**, 103–110 (2016b). https://doi.org/10.1016/j.solener.2016.05.046
65. H.A. Kazem, M.T. Chaichan, The effect of dust accumulation and cleaning methods on PV panels' outcomes based on an experimental study of six locations in northern Oman. Sol. Energy **187**, 30–38 (2019). https://doi.org/10.1016/j.solener.2019.05.036

66. J. Alonso-Montesinos, F.R. Martínez, J. Polo, N. Martín-Chivelet, F.J. Batlles, Economic effect of dust particles on photovoltaic plant production. Energies **13**(23), 6376 (2020). https://doi.org/10.3390/en13236376
67. H.M. Walwil, A. Mukhaimer, F.A. Al-Sulaiman, S.A.M. Said, Comparative studies of encapsulation and glass surface modification impacts on PV performance in a desert climate. Sol. Energy **142**, 288–298 (2017). https://doi.org/10.1016/j.solener.2016.12.020
68. E.M. Galal, A.S. Abdel-Mawgoud, M.H. Hamed, G.A. Ali, The performance of polycrystalline and monocrystalline solar modules under the climate conditions of el-kharga oasis, new valley governorate, Egypt. Int. J. Thin Film Sci. Technol. **12**(3), 207–215 (2023). https://doi.org/10.18576/ijtfst/120306
69. G.T. Chala, S.A. Sulaiman, S.M. Al Alshaikh, Effects of climatic conditions of Al Seeb in Oman on the performance of solar photovoltaic panels. Heliyon, e30944 (2024). https://doi.org/10.1016/j.heliyon.2024.e30944
70. V. Keskin, Energy- and exergy-based economical and environmental (4E) evaluation of the influence of natural pollutants on PV array performance. J. Therm. Anal. Calorim. (2024). https://doi.org/10.1007/s10973-024-13160-1
71. Y.A. Badamasi, S. Oodo, N.B. Gafai, F.B. Ilyasu, Effect of tilt angle and soiling on photovoltaic modules losses, in *2021 1st international Conference on Multidisciplinary Engineering and Applied Science (ICMEAS)*, (IEEE, 2021). https://doi.org/10.1109/icmeas52683.2021.9692375
72. A. Shariah, E. Al-Ibrahim, Impact of dust and shade on solar panel efficiency and development of a simple method for measuring the impact of dust in any location. J. Sustain. Dev. Energy Water Environ. Syst. **11**(2), 1–14 (2023). https://doi.org/10.13044/j.sdewes.d11.0448
73. M. Jaszczur, J. Teneta, K. Styszko, Q. Hassan, P. Burzyńska, E. Marcinek, N. Łopian, The field experiments and model of the natural dust deposition effects on photovoltaic module efficiency. Environ. Sci. Pollut. Res. **26**(9), 8402–8417 (2018). https://doi.org/10.1007/s11356-018-1970-x
74. A. Khodakaram-Tafti, M. Yaghoubi, Experimental study on the effect of dust deposition on photovoltaic performance at various tilts in semi-arid environment. Sustain Energy Technol Assess **42**, 100822 (2020). https://doi.org/10.1016/j.seta.2020.100822
75. D. Olivares, P. Ferrada, J. Bijman, S. Rodríguez, M. Trigo-González, A. Marzo, J. Rabanal-Arabach, J. Alonso-Montesinos, F.J. Batlles, E. Fuentealba, Determination of the soiling impact on photovoltaic modules at the coastal area of the Atacama desert. Energies **13**(15), 3819 (2020). https://doi.org/10.3390/en13153819
76. A.H. Shah, A. Hassan, M.S. Laghari, A. Alraeesi, The influence of cleaning frequency of photovoltaic modules on power losses in the desert climate. Sustainability **12**(22), 9750 (2020). https://doi.org/10.3390/su12229750
77. M.Z. Farahmand, M.E. Nazari, S. Shamlou, M. Shafie-khah, The simultaneous impacts of seasonal weather and solar conditions on PV panels electrical characteristics. Energies **14**(4), 845 (2021). https://doi.org/10.3390/en14040845
78. M. Mostefaoui, A. Ziane, A. Bouraiou, S. Khelifi, Effect of sand dust accumulation on photovoltaic performance in the Saharan environment: Southern Algeria (Adrar). Environ. Sci. Pollut. Res. **26**(1), 259–268 (2018). https://doi.org/10.1007/s11356-018-3496-7
79. S.C.S. Costa, S.d.M. Hanriot, A.S.A.C. Diniz, L.L. Kazmerski, C.B. Maia, C.D. Campos, D.S. Braga, P.P. Brito, V.C. Santana, E.M. Barboso, Comparative investigations of the effects of soiling of PV modules and systems in tropical, subtropical, and semi-arid climate zones in Brazil, in *2019 IEEE 46th Photovoltaic Specialists Conference (PVSC)*, (IEEE, 2019). https://doi.org/10.1109/pvsc40753.2019.8980505
80. W. Javed, B. Guo, B. Figgis, B. Aïssa, Dust potency in the context of solar photovoltaic (PV) soiling loss. Sol. Energy **220**, 1040–1052 (2021). https://doi.org/10.1016/j.solener.2021.04.015
81. M. Konyu, N. Ketjoy, C. Sirisamphanwong, Effect of dust on the solar spectrum and electricity generation of a photovoltaic module. IET Renewable Power Gen. **14**(14), 2759–2764 (2020). https://doi.org/10.1049/iet-rpg.2020.0456

82. M. Benghanem, S. Haddad, A. Alzahrani, A. Mellit, H. Almohamadi, M. Khushaim, M.S. Aida, Evaluation of the performance of polycrystalline and monocrystalline PV technologies in a hot and arid region: An experimental analysis. Sustainability **15**(20), 14831 (2023). https://doi.org/10.3390/su152014831
83. E. Abdeen, E.-S. Hasaneen, M. Orabi, Real study for Photovoltaic system performance in desert environment—Upper Egypt—case study, in *2016 Eighteenth International Middle East Power Systems Conference (MEPCON)*, (IEEE, 2016). https://doi.org/10.1109/mepcon.2016.7836993
84. I. Radonjić, T. Pavlović, D. Mirjanić, L. Pantić, Investigation of fly ash soiling effects on solar modules performances. Sol. Energy **220**, 144–151 (2021b). https://doi.org/10.1016/j.solener.2021.03.046
85. L.S. Alvarenga, M.B. Luchini, W.T. Guimarães, *Avaliação do desempenho da geração fotovoltaica na região metropolitana da grande Vitória diante da poluição atmosférica* (Anais Congresso Brasileiro De Energia Solar—CBENS, 2020). https://doi.org/10.59627/cbens.2020.759
86. D. Olivares, P. Ferrada, C.D. Matos, A. Marzo, E. Cabrera, C. Portillo, J. Llanos, Characterization of soiling on PV modules in the Atacama Desert. Energy Procedia **124**, 547–553 (2017). https://doi.org/10.1016/j.egypro.2017.09.263
87. W. Javed, Y. Wubulikasimu, B. Figgis, B. Guo, Characterization of dust accumulated on photovoltaic panels in Doha, Qatar. Sol. Energy **142**, 123–135 (2017). https://doi.org/10.1016/j.solener.2016.11.053
88. H. Zitouni, A. Azouzoute, C. Hajjaj, M. El Ydrissi, M. Regragui, J. Polo, A. Oufadel, A. Bouaichi, A. Ghennioui, Experimental investigation and modeling of photovoltaic soiling loss as a function of environmental variables: A case study of semi-arid climate. Sol. Energy Mater. Sol. Cells **221**, 110874 (2021). https://doi.org/10.1016/j.solmat.2020.110874
89. P. Ferrada, D. Olivares, V. del Campo, A. Marzo, F. Araya, E. Cabrera, J. Llanos, J. Correa-Puerta, C. Portillo, D. Román Silva, M. Trigo-Gonzalez, J. Alonso-Montesinos, G. López, J. Polo, F.J. Batlles, E. Fuentealba, Physicochemical characterization of soiling from photovoltaic facilities in arid locations in the Atacama Desert. Sol. Energy **187**, 47–56 (2019). https://doi.org/10.1016/j.solener.2019.05.034
90. R. Conceição, H.G. Silva, J. Mirão, M. Gostein, L. Fialho, L. Narvarte, M. Collares-Pereira, Saharan dust transport to Europe and its impact on photovoltaic performance: A case study of soiling in Portugal. Sol. Energy **160**, 94–102 (2018). https://doi.org/10.1016/j.solener.2017.11.059
91. A. Alraeesi, A.H. Shah, A. Hassan, M.S. Laghari, Characterisation of dust particles deposited on photovoltaic panels in The United Arab Emirates. Appl. Sci. **13**(24), 13162 (2023). https://doi.org/10.3390/app132413162
92. P.V.V. Romanholo, B.P. de Alvarenga, E.G. Marra, S.P. Pimentel, *Sujidade depositada sobre módulos fotovoltaicos instalados em Goiânia: morfologia e composição química* (Anais Congresso Brasileiro De Energia Solar—CBENS, 2018). https://doi.org/10.59627/cbens.2018.227
93. K. Ilse, M. Werner, V. Naumann, B.W. Figgis, C. Hagendorf, J. Bagdahn, Microstructural analysis of the cementation process during soiling on glass surfaces in arid and semi-arid climates. Phys. Status Solidi (RRL)—Rapid Res. Lett. **10**(7), 525–529 (2016). https://doi.org/10.1002/pssr.201600152
94. K.K. Ilse, B.W. Figgis, M. Werner, V. Naumann, C. Hagendorf, H. Pöllmann, J. Bagdahn, Comprehensive analysis of soiling and cementation processes on PV modules in Qatar. Sol. Energy Mater. Sol. Cells **186**, 309–323 (2018). https://doi.org/10.1016/j.solmat.2018.06.051
95. X. Li, D.L. Mauzerall, M.H. Bergin, Global reduction of solar power generation efficiency due to aerosols and panel soiling. Nat. Sustainability **3**(9), 720–727 (2020). https://doi.org/10.1038/s41893-020-0553-2
96. International Energy Agency Photovoltaic Power Systems Programme (IEA PVPS): Soiling losses—Impact on the performance of photovoltaic power plants (2022), https://iea-pvps.org/key-topics/soiling-losses-impact-on-the-performance-of-photovoltaic-power-plants/. Accessed 19 June 2025

Chapter 4
Experimental Procedures

4.1 Study Location

4.1.1 Environmental Characterization

Since environmental factors are strongly related to the study of soiling, it is necessary to characterize such aspects of the site. Hence, seasonal behaviors of ambient temperature, horizontal irradiation, relative humidity, and rainfall in the city of Fortaleza from 2012 to 2022 are analyzed. Rainfall data are obtained from a station of the Ceará Foundation for Meteorology and Water Resources (FUNCEME) [1]; the other environmental parameters come from an Automatic Station (A305) of the National Institute of Meteorology (INMET), approximately 9.5 km from the PV plant [2, 3]. Maximum and minimum value filters are used to eliminate measurement errors of ambient temperature and relative air humidity; these filters are defined based on the Brazilian Climatological Standards, developed by INMET, covering the period 1991–2020 [4]. Hence, ambient temperature measurements between 19.9 and 36.5 °C and relative air humidity values between 39.0 and 97.5% are used.

Fortaleza is characterized by a rainy season in the first semester of the year and a dry season in the second one, a typical meteorological condition of the Brazilian semi-arid. Monthly rainfall volumes (maximum, average, and minimum) and standard deviation for the period 2012 to 2022 [3] in the city are shown in Fig. 4.1; statistical data are specified in Table 4.1.

March is usually the rainiest month, with an average of 396.55 mm; September is usually the driest, with an average of 7.05 mm. A transition between the rainy period (February–May) and the dry period (August–November) is identified, corresponding to the post-rainy period (June and July) and pre-rainy period (December

J. J. Silva de Souza, P. C. Marques de Carvalho, *Soiling Effects on Photovoltaic Systems*, SpringerBriefs in Applied Sciences and Technology,
https://doi.org/10.1007/978-3-032-22442-2_4

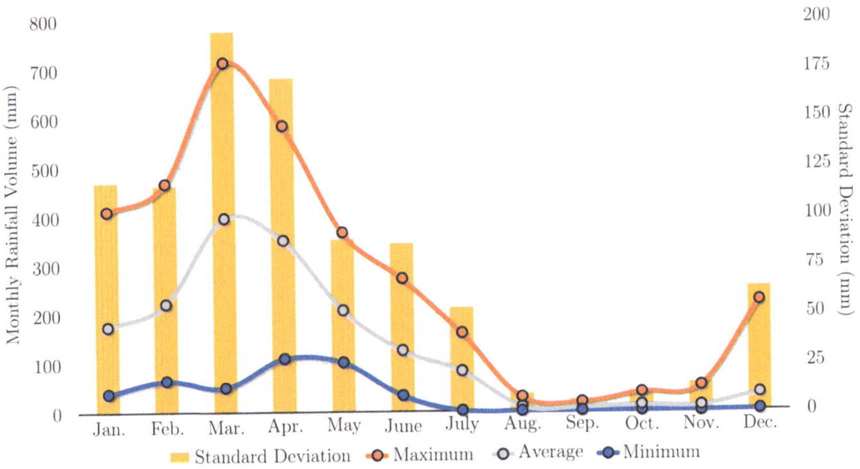

Fig. 4.1 Monthly rainfall profile in Fortaleza (2012–2022)

Table 4.1 Statistical data of monthly rainfall in Fortaleza (2012–2022)

Metrics (mm)	Jan.	Feb.	Mar.	Apr.	May	June	July	Aug.	Sep.	Oct.	Nov.	Dec.
Standard deviation	117.8	116.0	194.8	171.0	88.6	86.5	53.4	9.4	6.3	10.7	14.7	63.6
Minimum	38.3	64.6	50.3	109.7	101.1	35.6	1.1	0.0	0.0	0.0	0.0	1.3
Maximum	410.3	466.7	714.1	584.3	366.6	271.7	160.1	28.4	17.4	37.8	51.0	223.7
Average	175.1	221.5	396.5	350.7	207.4	125.0	82.5	9.0	7.0	10.9	9.6	35.5

and January). Analyzing the standard deviation, a relationship similar to total rainfall is observed: months with higher rainfall have larger standard deviations, highlighting the irregularity of rainfall over time, as reported by [5], who analyzed rainfall records spanning 62 years (1951–2012) in Fortaleza.

During the second semester, the dry period in Fortaleza, the highest levels of solar irradiation are recorded: in August (5.43 kWh/m^2/day), September (5.78 kWh/m^2/day), October (5.86 kWh/m^2/day), and November (5.67 kWh/m^2/day); the lowest levels occur in March (4.47 kWh/m^2/day), April (4.44 kWh/m^2/day), and May (4.52 kWh/m^2/day); the average global horizontal irradiation is 5.04 kWh/m^2/day. Monthly values of global horizontal irradiation in the city are shown in Fig. 4.2.

Considering the period 2012–2022, ambient temperature data show an annual average of 27.3 °C; the highest temperature recorded is 34.4 °C in March; the lowest is 20.1 °C in July. The ambient temperature profile in Fortaleza in the period is shown in Fig. 4.3; statistical data are specified in Table 4.2. Despite the low variation of the average monthly temperature, the standard deviation is greater in the post-rainy period (June and July) and smaller in the pre-rainy period (December and January).

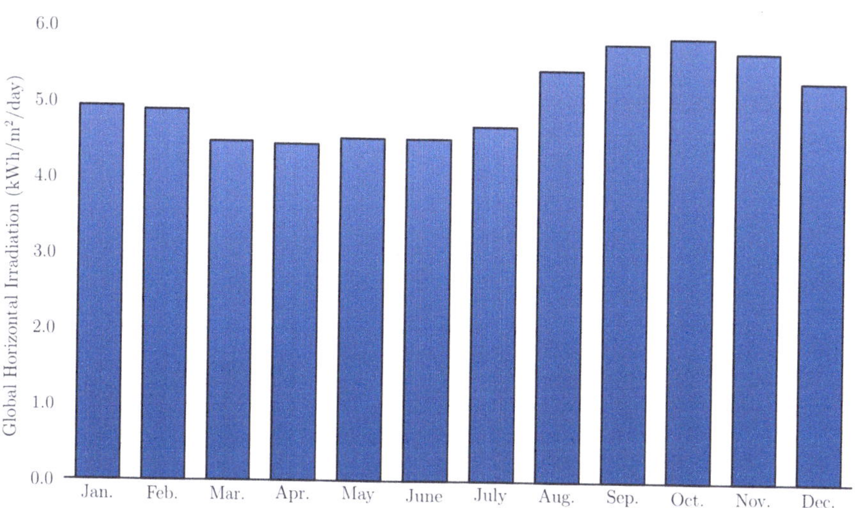

Fig. 4.2 Monthly global horizontal irradiation in Fortaleza (2012–2022)

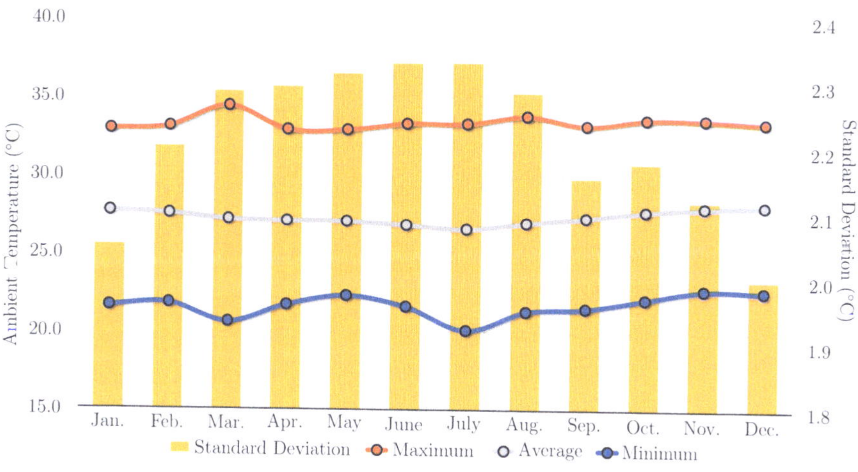

Fig. 4.3 Monthly profile of ambient temperature in Fortaleza (2012–2022)

The monthly profile of the relative air humidity in Fortaleza in the period is shown in Fig. 4.4; statistical data are specified in Table 4.3: the highest values are found in the first semester, coinciding with the rainy period. Considering the rainy and dry seasons, the averages are 76.8% and 66.3%, respectively; the annual average is 71.4%.

Analyzing the standard deviation of the relative air humidity, highest values are recorded in June (12.7%) and July (12.9%), the transition period between rainy and dry periods. Hence, the local relative air humidity presents stable behavior throughout the year.

Table 4.2 Statistical data of monthly ambient temperature in Fortaleza (2012–2022)

Metrics (°C)	Jan.	Feb.	Mar.	Apr.	May	June	July	Aug.	Sep.	Oct.	Nov.	Dec.
Standard deviation	2.0	2.2	2.2	2.3	2.3	2.3	2.3	2.2	2.1	2.1	2.1	2.0
Minimum	21.6	21.8	20.6	21.7	22.3	21.6	20.1	32.3	21.5	22.1	22.7	22.6
Maximum	36.9	33.1	34.4	32.9	32.9	33.3	33.3	33.8	33.2	33.6	33.6	33.4
Average	27.6	27.5	27.1	27,0	27.0	26.8	26.5	26.9	27.3	27.7	27.9	28.1

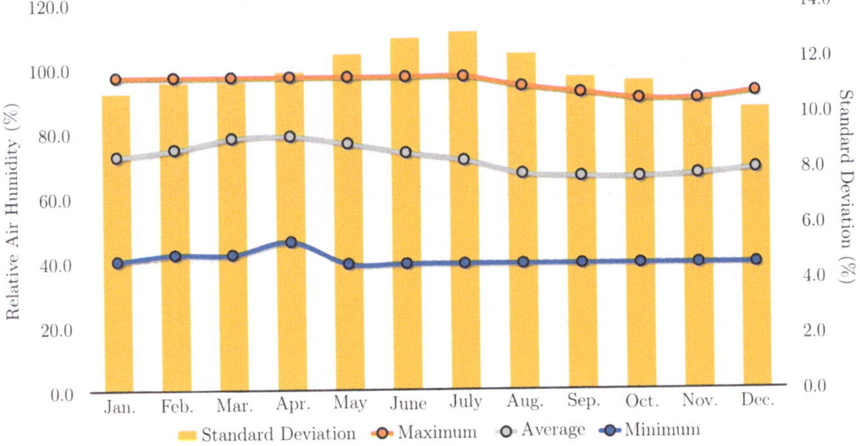

Fig. 4.4 Monthly profile of relative air humidity in Fortaleza (2012–2022)

Table 4.3 Statistical data of monthly relative air humidity in Fortaleza (2012–2022)

Metrics (%)	Jan.	Feb.	Mar.	Apr.	May	June	July	Aug.	Sep.	Oct.	Nov.	Dec.
Standard deviation	10.7	11.1	11.1	11.5	12.1	12.7	12.9	12.1	11.3	11.1	10.5	10.1
Minimum	40.0	42.0	42.0	46.0	39.0	39.0	39.0	39.0	39.0	39.0	39.0	39.0
Maximum	97.0	97.0	97.0	97.0	97.0	97.0	97.0	94.0	92.0	90.0	90.0	92.0
Average	72.3	74.5	78.0	78.6	76.3	73.3	71.0	66.8	65.9	65.7	66.6	68.2

4.1.2 *LEA2 PV Plant Description*

The PV plant used in the study, called LEA2 PV plant, has a nominal power of 3.9 kWp, composed of 12 330 Wp p-Si modules, distributed in two strings connected in parallel (six modules in series in each string). The plant has a 10° inclination, is oriented to the geographic north, and is installed in a free span. An aerial view of the plant site is shown in Fig. 4.5, highlighting a busy avenue in the neighborhood; LEA2 modules technical specifications are listed in Table 4.4. LEA2 uses a PHB5000D-NS inverter for grid connection, with fuses and surge protection devices; the datalogger, integrated with the inverter, has a Wi-Fi connection,

Fig. 4.5 Aerial view of LEA–UFC PV plants. The designations ST1 and ST2 in this figure represent strings 1 and 2 of the LEA2 plant, respectively

Table 4.4 Specifications of the LEA2 modules in STC [6]

Parameter	Specifications
Manufacturer	Jinko Solar
Model	JKM330PP-72
Maximum power (P_{Max})	330 W
Module efficiency (η)	17.01%
Maximum power voltage (V_{Max})	37.80 V
Maximum power current (I_{Max})	8.74 A
Open-circuit voltage (V_{oc})	46.90 V
Short-circuit current (I_{sc})	9.14 A
Temperature coefficients of P_{Max} (γ)	−0.38%/°C
Temperature coefficients of V_{oc} (β_{Voc})	−0.31%/°C
Temperature coefficients of I_{sc} (α)	−0.06%/°C
Dimensions (L × W × H)	1956 mm × 992 mm × 40 mm

Table 4.5 LP02 pyranometer specifications [8]

Parameter	Specifications
ISO classification (ISO 9060:1990)	Second class
Response time (95%)	18 s
Measurement range	0–2000 W/m^2
Sensitivity range	7–25·10^{-6} V/(W/ m^2)
Sensitivity (nominal)	5·10^{-6} V/(W/m^2)
Rated operating temperature range	−40 to +80 °C
Spectral range (20% transmission points)	285–3000 10^{-9} m

Table 4.6 NRG #40C anemometer specifications [9]

Parameter	Specifications
Sensor range	1–96 m/s
Output signal range	0–125 Hz
Threshold	0.78 m/s
Operating humidity range	0–100%
Operating temperature range	−55–60 °C

Table 4.7 DHT11 sensor specifications [10]

Parameter	Specifications
Sensor range	20–90% and 0–50 °C
Accuracy (humidity)	±5%
Accuracy (temperature)	±2 °C

recording electrical parameters; a sampling rate of 1 min is used, with data being sent and stored on the inverter manufacturer's web server.

4.1.3 Monitoring and Acquisition of Environmental and Operational Data

An IoT (Internet of Things)-based system, proposed by [7], is used to monitor environmental and operational conditions; the system is adapted to send data via Wi-Fi. A ThingSpeak server is used to store and provide the collected data.

Irradiance data is acquired with an LP02 pyranometer (Table 4.5), wind speed is measured with an NRG #40C anemometer (Table 4.6), and relative air humidity is measured with a DHT11 sensor (Table 4.7). The sampling rate adopted for all measurements is 1 min.

Digital sensors (model DS18B20) are used for T_m monitoring, which have waterproof encapsulation and three terminals (GND—Ground; DATA—Data output and VDD—Power supply); additional information is presented in Table 4.8.

Initially, the sensors are subjected to a calibration process adopting immersion procedures in temperature standards and reference instruments. Tests are performed

Table 4.8 DS18B20 sensor characteristics [11]

Parameter	Specifications
Sensor range	−55 to 125 °C
Supply voltage	3–5.5 V
Accuracy	±0.5 °C (−10 to 85 °C) and ± 2 °C (−55 to 125 °C)
Conversion time	750 ms
Resolution	9 bits

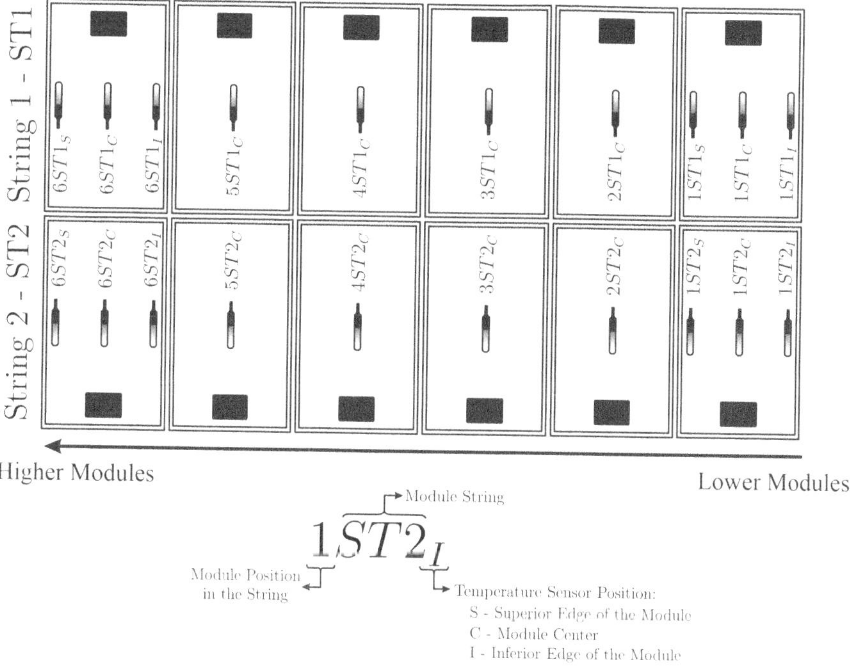

Fig. 4.6 LEA2 plant rear side with temperature sensors identification

at three temperatures: (1) 0 °C (solid/liquid equilibrium—insulated container with ice); (2) 24.50 °C—ambient temperature of the location, using a mercury thermometer as a reference, and (3) 100 °C (liquid/gas equilibrium—container with water boiler).

To collect operational information from the LEA2 PV plant, the temperature is monitored at the center of all modules and at the upper and lower edges of the modules at the ends of the PV plant. Sensors' position and their respective nomenclatures are shown in Fig. 4.6.

To ensure accurate measurements, thermal paste was applied to promote effective thermal contact between the sensors and the module backsheet. Sequentially, the sensors were fixed using a high-temperature silicone adhesive, while polyethylene foam with a protective film was employed as thermal insulation to mitigate

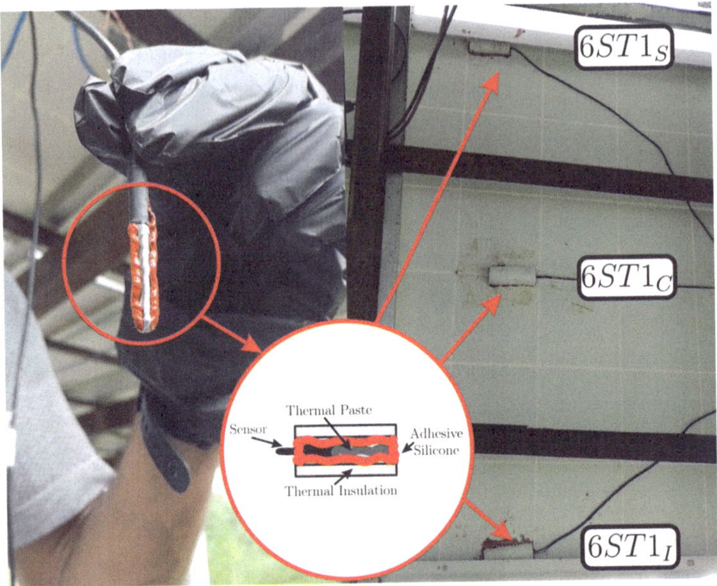

Fig. 4.7 Temperature sensors installation on the backside of LEA2's modules

Table 4.9 Estimated uncertainty values of the LEA2 measuring instruments

Parameter	Uncertainty (%)
Irradiance (LP02)	1.80
Wind speed (NRG #40C)	1.48
Module temperature (DS18B20)	1.00 [13]
Ambient temperature (DS18B20)	1.00 [13]
Relative humidity (DHT11)	1.00 [14]
Electrical (PHB5000D-NS)	1.00 [15]
Total	**2.75**

environmental interference in the data acquisition process. The sensors installation procedure is presented in Fig. 4.7.

The uncertainty levels, estimated as proposed by [12], including electrical parameters (measured directly by the inverter) and meteorological variables (monitored by the integrated IoT system), are shown in Table 4.9.

4.2 Manual Cleaning

An overview of the LEA2 plant, prior to the manual cleaning performed on July 5, 2023, is shown in Fig. 4.8. A uniformly distributed layer of soiling is evident, along with the presence of excrement, particularly on the lower modules, where incrustations hinder natural removal.

Fig. 4.8 LEA2 plant before the first cleaning

Fig. 4.9 Steps for performing manual cleaning of LEA2 modules

A 3.0 m window-cleaning mop, mild detergent, potable water, microfiber cloths, bucket, and water sprayer were employed for the manual cleaning. The procedure was carried out in the morning to favor optimal conditions, with the use of mild detergent to facilitate the removal of the encrusted soiling. The cleaning process consisted of the following steps: (1) wetting the modules; (2) scrubbing the modules with the mop moistened in a detergent solution; (3) rinsing the modules with water; (4) removing excess water; and (5) drying the modules with a microfiber cloth. When necessary, the procedure was repeated. The cleaning sequence applied to the LEA2 modules is illustrated in Fig. 4.9.

Considering the plant installation configuration, the cleaning process is conducted in two stages: (1) starting from the plant top, modules 4, 5, and 6 of each string are cleaned; and (2) using ladders from the base of the plant, modules 1, 2, and 3 are subsequently cleaned. An overview of the plant before, during, and after the manual cleaning performed on July 5, 2023, is shown in Fig. 4.10.

As the plant is divided into two strings (ST1 and ST2), each connected to a dedicated inverter MPPT (MPPT1 and MPPT2), voltage and current measurements are

Fig. 4.10 LEA2 plant before, during, and after manual cleaning on July 5th, 2023

Table 4.10 Schedule for manual cleaning (MC) of LEA2 PV modules

Year	Date	Count
2023	July 5, 12, 19, 26; August 2, 16, 30; September 13, 27; October 18; November 8, 29; and December 20	13
2024	January 10, 17, 24, 31; February 7; March 6; April 3; May 8; June 12; July 17; August 7, 14, 28; September 18; October 16; November 20; and December 11	17
2025	January 15	1

used to determine the power output of each string. Following the intervention on July 5th, 2023, only the modules in the ST2 string are systematically subjected to manual cleaning at varying intervals, aiming to evaluate the soiling impact on the electrical performance; ST1 modules remain exposed only to natural cleaning. Manual cleaning dates are listed in Table 4.10; from this point onward, the notation "MC" followed by a sequential number is adopted to identify each manual cleaning event.

4.2.1 *Manual Cleaning Effects on Electrical Performance*

The analysis, from January 2023 to January 2025, is structured into three phases:

- Phase I (January 4–July 4, 2023): both ST1 and ST2 operated without manual cleaning.
- Phase II (July 5–July 11, 2023): both ST1 and ST2 underwent manual cleaning.
- Phase III (July 12, 2023–January 28, 2025): only ST2 was subjected to manual cleaning, with intervals ranging from 1 to 5 weeks.

To mitigate interruptions effects on the measurement of electrical parameters, the following criteria were adopted:

- Only days with at least 75% of the expected data available between 10:00 and 14:00 (peak generation period) and between 07:00 and 17:00 (full LEA2 operation) are considered.
- The interval from 07:00 to 17:00 is used for the calculation of the total daily generation.

Initially, the relative generation percentage difference ($\%E_{LEA2}$), calculated according to Eq. (4.1), is used to evaluate soiling losses in LEA2; E_{ST_1} and E_{ST_2} represent ST_1 and ST_2 generation, respectively.

$$\%E_{LEA2} = \left(1 - \frac{E_{ST_1}}{E_{ST_2}}\right) \bullet 100\% \tag{4.1}$$

For the comparative analysis between rainfall regimes, with and without cleaning, the metrics Yield, CF, and PR, in addition to $\%E_{LEA2}$, were employed, as defined in Eqs. (4.2)–(4.4). To prevent distortions in the behavior of these metrics, only days with irradiation levels greater than or equal to 2.50 kWh/m^2 were considered.

$$\text{Yield} = \frac{E_{ST_{1-2}}}{P_{STC}} \tag{4.2}$$

$$CF = \frac{E_{ST_{1-2}}}{24 \cdot P_{STC}} \tag{4.3}$$

$$PR = \frac{E_{ST_{1-2}}}{P_{STC} \cdot \frac{H_{POA}}{G_{STC}}} \tag{4.4}$$

Based on these premises, the execution steps for this study stage are shown in Fig. 4.11.

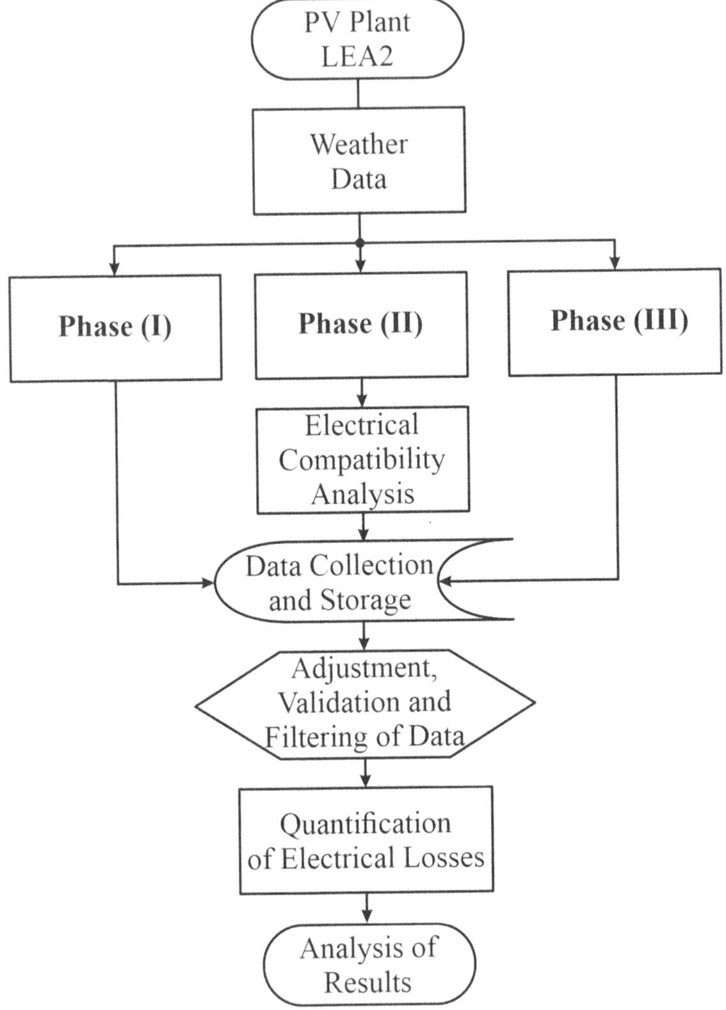

Fig. 4.11 Flowchart of the study of manual cleaning impacts on LEA2 generation

References

1. FUNCEME—Ceará Foundation for Meteorology and Water Resources: Postos pluviométricos. FUNCEME (2022), http://www.funceme.br/?page_id=2694
2. INMET—National Institute of Meteorology: Mapa de estações. National Institute of Meteorology (2022a), https://mapas.inmet.gov.br/
3. INMET—National Institute of Meteorology: Dados históricos anuais. National Institute of Meteorology (2025), https://portal.inmet.gov.br/dadoshistoricos
4. INMET—National Institute of Meteorology: Normais climatológicas do Brasil. National Institute of Meteorology (2022b), https://portal.inmet.gov.br/normais

5. Í.J.M. Moura, D.F.d. Santos, F.G.d.M. Pinheiro, C.J.d. Oliveira, Caracterização dos períodos seco e chuvoso da cidade de Fortaleza (CE). Ciência e Natura **37** (2015). https://doi.org/10.5902/2179460x16206
6. Jinko Solar: JKM330PP-72 310–330 watt (n.d.), www.jinkosolar.com. https://www.enf.com.cn/Product/pdf/Crystalline/57aaf04f4a043.pdf
7. D.P.D. Assis, L.D.O. Santos, R.I.S. Pereira, J.J.S. De Souza, P.C.M.D. Carvalho, Scalable data acquisition system for real-time monitoring of photovoltaic plants. IEEE Access **1** (2025). https://doi.org/10.1109/access.2025.3609654
8. HUKSEFLUX: USER MANUAL LP02 second class pyranometer. Hukseflux Thermal Sensors (2008), https://www.hukseflux.com/uploads/inline/LP02_manual_v2008_DISCONTINUED.pdf
9. NRG Systems: NRG 40C anemometer. NRG Systems—Wind + Solar measurement tools (2010), https://www.nrgsystems.com/products/met-sensors/detail/40c-anemometer
10. Mouser Electronics: DHT11 humidity & temperature sensor (2021), https://www.mouser.com/datasheet/2/758/DHT11-Technical-Data-Sheet-Translated-Version-1143054.pdf
11. R. Lousada, Guia completo do sensor DS18B20 a prova d'água. Blog Eletrogate (2020), https://blog.eletrogate.com/guia-completo-sobre-sensor-de-temperatura-ds18b20-a-prova-dagua/?_gl=1*1lnqgyq*_gcl_au*ODQ4ODkxNjIyLjE3NjQ1OTIyOTE
12. M. Al-Housani, Y. Bicer, M. Koç, Experimental investigations on PV cleaning of large-scale solar power plants in desert climates: Comparison of cleaning techniques for drone retrofitting. Energy Convers. Manag. **185**, 800–815 (2019). https://doi.org/10.1016/j.enconman.2019.01.058
13. D. Yulizar, S. Soekirno, N. Ananda, M.A. Prabowo, I.F.P. Perdana, D. Aofany, Performance analysis comparison of DHT11, DHT22 and DS18B20 as temperature measurement, in *Proceedings of the 2nd International Conference on Science Education and Sciences 2022 (ICSES 2022)*, (Atlantis Press International BV, 2023), pp. 37–45. https://doi.org/10.2991/978-94-6463-232-3_5
14. V. Gupta, M. Sharma, R. Pachauri, K.N.D. Babu, Performance analysis of Solar PV system using customize wireless data acquisition system and novel cleaning technique. Energy Sources Part A: Recovery Util. Environ. Effects **44**(2), 2748–2769 (2022). https://doi.org/10.1080/15567036.2022.2061091
15. A. Arabkoohsar, Dynamic modeling of a compressed air energy storage system in a grid connected photovoltaic plant, (PhD in Mechanical Engineering), Federal University of Minas Gerais, 2016

Chapter 5
Experimental Results

5.1 Effects of Manual Cleaning on PV Electrical Performance

The weekly behavior of $\%E_{LEA2}$ is shown in Fig. 5.1, with black and orange circles marking the instances of manual cleaning; week 9 is excluded from the analysis due to interruptions in environmental and electrical measurements. To enable clear correlation between weekly intervals and study phases, the week in which both strings were cleaned is designated as Week 0 (Phase II), serving as a reference point for the preceding (Phase I) and following (Phase III) weeks. Weeks prior to this full cleanup are numbered from −26 to −1, while weeks in which only ST2 was cleaned are numbered from 1 to 81.

During Phase I (Week −26 to Week −1), ST1's electricity generation was, on average, 0.59% higher than that of ST2. Over this period, ST1 produced a total of 1357.21 kWh, while ST2 generated 1349.54 kWh, resulting in a difference of 7.68 kWh. The average daily generation was 7.62 kWh for ST1 and 7.58 kWh for ST2.

On July 5, 2023, LEA2 (MC0) was cleaned, and both strings were subjected to the same conditions until July 11, 2023 (Phase II—Week 0). During this week, $\%E_{LEA2}$ was 0.01%, a negligible value that demonstrates the electrical compatibility of ST1 and ST2 under identical conditions.

In Phase III (Week 1 to Week 81), ST2 generated a total of 4674.83 kWh, exceeding ST1's 4586.54 kWh by 88.29 kWh. Notably, in Week 3 (July 26 to August 1, 2023), after MC3, $\%E_{LEA2}$ reached 2.69% due to repair works on one of the lanes of a nearby avenue, which increased the accumulation of soiling on ST1 modules, as shown in Fig. 5.2.

Examining the daily behavior of LEA2 before and after MC3, an increase in $\%E_{LEA2}$ was recorded from 0.32% (1 day before MC3—July 25, 2023) to 2.79%

J. J. Silva de Souza, P. C. Marques de Carvalho, *Soiling Effects on Photovoltaic Systems*, SpringerBriefs in Applied Sciences and Technology,
https://doi.org/10.1007/978-3-032-22442-2_5

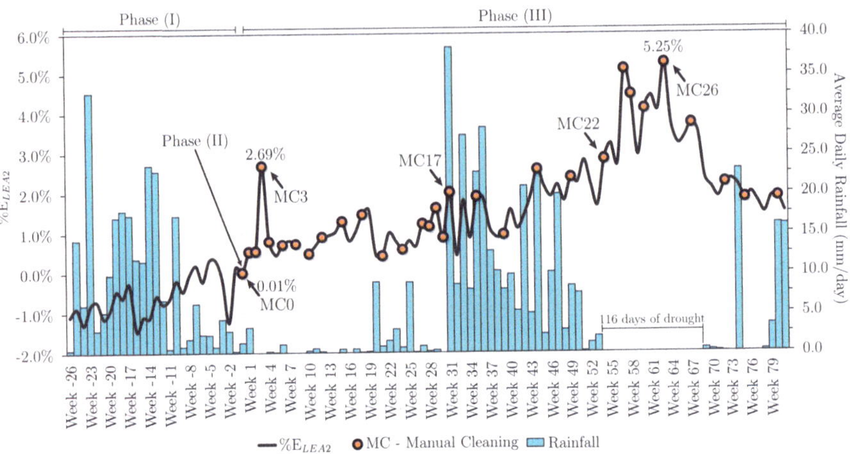

Fig. 5.1 Weekly behavior of $\%E_{LEA2}$

(1 day after MC3—July 27, 2023), despite an irradiation reduction of 7.75%; $\%E_{LEA2}$ increase was due to a layer of soiling on ST1 modules surface, a consequence of the repair works combined with the lack of manual cleaning and rainfall (last precipitation on July 18, 2023—9.60 mm). This event highlights the impact of human activities on PV modules soiling, especially during dry periods. Therefore, areas with regular pollutant/soiling emissions should be carefully analyzed when implementing PV plants, as short-term interventions have proven effective in reducing generation. In Week 4, a rainfall of 1.20 mm caused a sharp decrease in $\%E_{LEA2}$, from 2.69% to 0.79%. Between Weeks 5 and 19 (dry period of 2023), $\%E_{LEA2}$ gradually increased, ranging from 0.44% to 1.59%.

Between Weeks 20 and 30, the end of the dry period and the pre-rainy period of 2023, increased rainfall (an average of up to 8.86 mm/day) resulted in sharp variations of $\%E_{LEA2}$, with a minimum of 0.44% (Week 20) and a maximum of 1.62% (Week 29). Although rainfall helps to mitigate soiling effects, it does not fully restore ST1 modules generation performance. Between Weeks 31 and 55, during the rainy and post-rainy period of 2024, maximum rainfall volume reached 38.31 mm/day, causing a variation of $\%E_{LEA2}$. MC17 (Week 31) increased $\%E_{LEA2}$ from 0.73% to 3.56%. The last rainfall in the period was recorded on January 29, 2024 (2.20 mm), while the highest recorded rainfall was 215.10 mm (February 1, 2024), contributing to the highest average daily rainfall in Week 31. Despite these rainfall events, $\%E_{LEA2}$ showed an increasing trend, indicating that natural cleaning is insufficient to fully restore the soiled modules performance. ST1 modules remained opaque (see Fig. 5.3), demonstrating rainfall limitation in completely removing the soiling. Furthermore, as reported in the literature, soiling losses are amplified under low irradiance conditions.

During the dry period of 2024 (Weeks 56 to 72), following MC26, $\%E_{LEA2}$ peaked at 5.25% in Week 63; this is the driest period of the study, with a total accumulated rainfall of 5.80 mm, illustrating that soiling intensifies during such periods, as

Fig. 5.2 Clean and dirty modules during repair works on an avenue near LEA2

shown in Fig. 5.4. After 15 weeks without significant rainfall, a 3.20 mm precipitation in Week 69 reduced $\%E_{LEA2}$ from 3.53% to 2.37%. From Weeks 73 to 81, $\%E_{LEA2}$ showed more stable behavior, with a downward trend due to increased rainfall, which reduced the soiling accumulation on the modules, contributing to a more balanced ST1 and ST2 electrical performance. These observations highlight the importance of considering precipitation dynamics in PV plants maintenance strategies.

Soiling is inherently a cumulative phenomenon, resulting in generation losses that tend to intensify in the absence of mitigation actions. In such context, the cumulative difference between ST2 and ST1 generation, in absolute terms, is shown in Fig. 5.5: negative values indicate that ST1 generated more than ST2, while positive values indicate the opposite. During Phase I, the cumulative difference showed a negative trend, decreasing from −0.65 kWh in Week −26 to −7.68 kWh in Week −1 (June 28 to July 4, 2023, before MC0). In Phase II, the cumulative difference remained essentially stable, varying by only 0.01 kWh. In Phase III, the cumulative difference increased positively, with ST2 producing 88.29 kWh more than ST1; considering ST2 average daily generation of 7.79 kWh during this phase, this difference corresponds to approximately 11.33 days of ST2 operation.

Fig. 5.3 LEA2 modules after MC22 (July 17th, 2024)

$\%E_{LEA2}$ monthly behavior is shown in Fig. 5.6: in January 2023, ST2 generated approximately 1% less than ST1; however, with the implementation of the cleaning schedule in July 2023, ST2 generation began to exceed that of ST1, reaching a maximum $\%E_{LEA2}$ of 4.44% in September 2024. The transition from the rainy period

Fig. 5.4 Clean and dirty LEA2 modules during the dry period after MC26

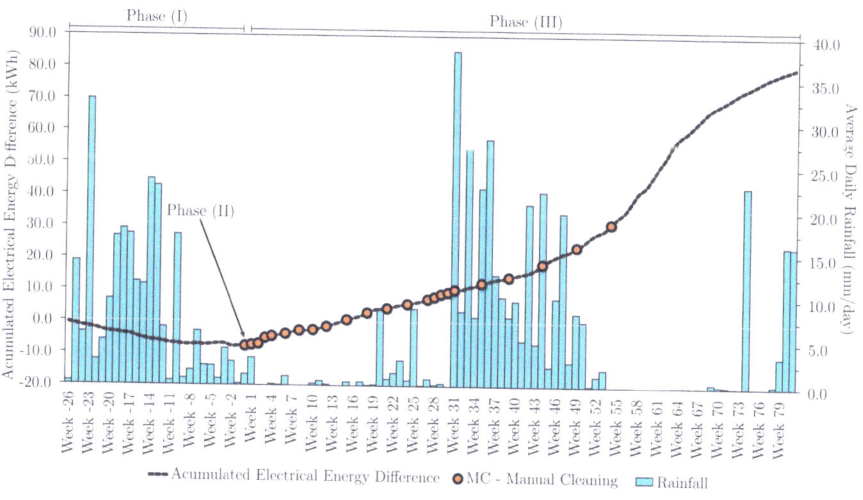

Fig. 5.5 Cumulative difference of electricity generation of ST2 compared to ST1

to the post-rainy period, between May and June 2023, is characterized by a gradual increase of the irradiation levels. Frequent rainfall during this period helped wash ST1 and ST2 modules, reducing soiling and enabling the strings to generate at similar levels. Furthermore, the following aspects are noteworthy:

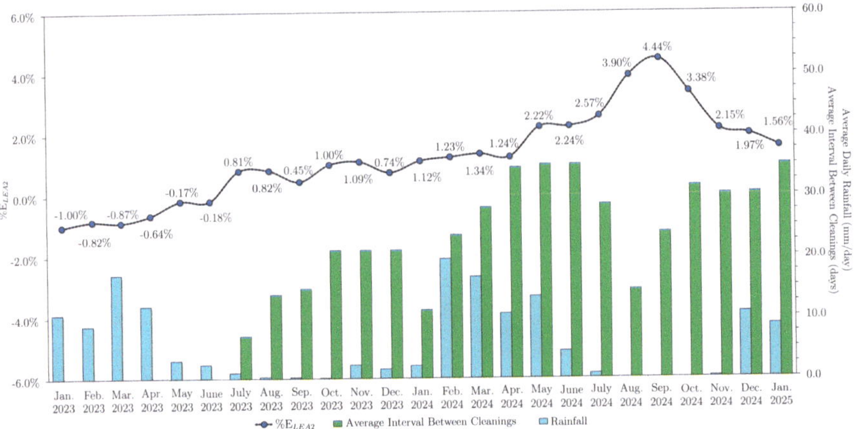

Fig. 5.6 Monthly behavior of $\%E_{LEA2}$ as a function of precipitation volume and average interval between manual cleanings

- **July–August 2023**: Despite an increase of the average interval between cleanings from 7 days to 11.67 days and a decrease of the average daily rainfall from 1 mm/day to 0.30 mm/day, $\%E_{LEA2}$ remained stable (0.81% to 0.82%), indicating that soiling is effectively controlled due to the relatively short cleaning intervals.
- **September 2023**: The lowest $\%E_{LEA2}$ value (0.45%) corresponded to an average rainfall volume of 0.24 mm/day and an average cleaning interval of 14 days. The combination of regular manual cleanings and light rainfall proved sufficient to maintain low soiling levels. A similar pattern was recorded from October to December 2023, where the average cleaning interval remained at 21 days, and the lowest $\%E_{LEA2}$ value (0.74%) was associated with an average rainfall volume of 1.55 mm/day.
- **January to June 2024**: $\%E_{LEA2}$ showed an upward trend, increasing from 1.12% to 2.24%, even though the average cleaning interval varied from 11.20 days to 35 days. Contributing factors include: (1) average daily rainfall varied between 2.08 mm/day and 19.50 mm/day, which helped to reduce soiling accumulation. However, due to the extended exposure of ST1 modules without manual cleaning, soiling gradually intensified, limiting the ability of rainfall to fully clean the modules, keeping ST1 consistently dirtier than ST2; (2) bird droppings deposits were only recorded on ST1 modules, as shown in Fig. 5.7. This irregular contamination increases soiling and can contribute to the formation of hot spots.
- **July to September 2024**: $\%E_{LEA2}$ increased markedly from 2.54% to 4.44%, despite an average interval reduction between cleanings from 28.47 to 23.90 days. This period was characterized by a lack of rainfall, intensifying generation losses. Although the period without rainfall extended into October 2024, the increase of the average cleaning interval to 31.50 days contributed to a decline of ST2 performance, bringing it closer to ST1. This trend highlights the pronounced soiling impact, exacerbated by the lack of natural cleaning.

Fig. 5.7 ST1 modules with bird droppings

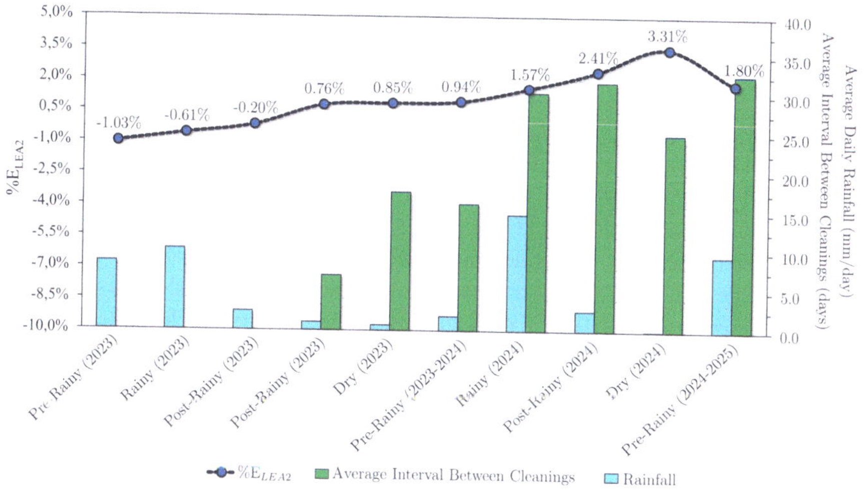

Fig. 5.8 $\%E_{LEA2}$ behavior as a function of average precipitation and intervals between cleanings

- **October 2024 to January 2025**: $\%E_{LEA2}$ generally decreased, while the average cleaning interval remained relatively stable (30.17–35 days). During this period, rainfall records, including a peak of 111.30 mm on December 5, 2024, contributed to the natural cleaning of the modules, reducing soiling losses.

$\%E_{LEA2}$ as a function of the average daily precipitation, cleaning intervals, and rainfall periods at the study site is shown in Fig. 5.8.

Between the post-rainy and pre-rainy periods of 2023, variations of the cleaning interval from 7 days to 17.64 days kept $\%E_{LEA2}$ relatively stable at around 0.85%. On average, each cleaning increased 0.58, 1.18, and 1.17 kWh in ST2 generation compared to ST1 in the post-rainy, dry, and pre-rainy periods of 2023, respectively. In 2024, the average cleaning interval ranged from 25.05 days (dry period) to 32.70 days (pre-rainy period), while $\%E_{LEA2}$ ranged from 1.57% (rainy period) to 3.51% (dry period). The highest $\%E_{LEA2}$ value was recorded during the dry period, when minimal average daily rainfall (0.05 mm/day) combined with the longest cleaning interval resulted in a significant soiling accumulation on the modules. Conversely, the lowest $\%E_{LEA2}$ value (1.57%) occurred during the rainy period, when the highest average daily rainfall (14.84 mm/day) and a cleaning interval of 30.33 days favored modules natural cleaning, minimizing soiling losses. These results reinforce the importance of rainfall as a determining factor in reducing soiling losses, demonstrating that periods with higher rainfall provide more effective natural cleaning and reduce the need for manual interventions.

Conversely, during prolonged periods without rainfall, the lack of natural cleaning exacerbates soiling effects, necessitating shorter intervals between manual cleanings to mitigate generation losses and maintain PV plant performance. Yield, CF, and PR metrics for the different rainfall regimes, with and without manual cleaning, are compared in Table 5.1.

Comparing the pre-rainy period of 2023 (without MC) to the pre-rainy period of 2023–2024 (with MC), all performance metrics for both strings showed positive gains. However, comparing the pre-rainy periods of 2023–2024 and 2024–2025 (both with MC), a downward trend of the metrics is identified. This behavior is primarily attributed to two factors: an average interval increases between cleanings from 16.18 days to 32.70 days, raising soiling accumulation, and an increase of the average daily rainfall from 1.85 mm/day to 9.74 mm/day. Although rainfall contributes to natural cleaning, it is insufficient for a complete soiling removal.

During rainy periods, it was demonstrated that, while rainfall acts as a natural mitigation agent, it cannot fully restore soiled modules performance. This was evidenced by a reduction of –0.25% in $Yield_{ST1}$ and –0.22% in CF_{ST1}. In contrast, clean modules showed opposite behavior, registering gains of 1.71% in $Yield_{ST2}$ and 1.76% in CF_{ST1}, indicating superior performance compared to ST1. PR values reinforced this trend: dirty modules showed a gain of 3.19% in PR_{ST1}, while ST2 increased by 5.21%, approximately 63.33% higher than ST1. This demonstrates that the combination of manual cleaning and regular rainfall enhances ST2 module performance.

Comparing the post-rainy period of 2023 (without MC) to the same period (with MC), ST1 and ST2 showed gains of 15.02% and 15.87% in Yield and CF, respectively, with ST1's gain approximately 5.35% lower than ST2. For the PR metric, ST2's gain (3.23%) is 48.85% higher than that of ST1 (2.17%). In contrast, comparing the post-rainy periods of 2023 and 2024, (both with MC), a reduction of Yield and CF gains for both strings was identified, though this reduction was less pronounced for ST2. PR_{ST2} again stands out, with an increase of 10.58%, 7.96% higher than that of ST1 (9.80%), reinforcing that ST2 benefits from regular cleaning, even with the increase of the average interval between manual interventions.

During the dry period, all ST1 metrics showed a reduction, notably a 3.37% reduction of PR_{ST1} when comparing the dry seasons of 2024 and 2023. This reduction is associated with the long period without rainfall recorded in the 2024 dry season (116 days of drought—last rainfall on July 17, 2024: 7.40 mm) and an

Table 5.1 Performance metrics for different rainfall regimes; the values for periods with MC are used as a reference to calculate the gain (%)

Rainfall period	$Yield_{ST1}$ (kWh/kWp)	$Yield_{ST2}$ (kWh/kWp)	CF_{ST2} (%)	CF_{ST2} (%)	PR_{ST1}	PR_{ST2}
Pre-rainy 2023 (without MC)	4.07	4.03	17.95	17.78	0.84	0.83
Pre-rainy 2023–2024 (with MC)	4.22	4.26	18.69	18.86	0.97	0.98
Pre-rainy 2024–2025 (with LM)	4.10	4.17	18.16	18.45	0.96	0.98
Gain (%) (without-with MC)	**3.55**	**5.40**	**3.96**	**5.73**	**13.40**	**15.31**
Gain (%)[a] (with MC)	**−2.93**	**−2.16**	**−2.92**	**−2.22**	**−1.04**	**0.00**
Rainy 2023 (without MC)	4.05	4.03	17.92	17.82	0.91	0.91
Rainy 2024 (with MC)	4.04	4.10	17.88	18.14	0.94	0.96
Gain (%) (without-with MC)	**−0.25**	**1.71**	**−0.22**	**1.76**	**3.19**	**5.21**
Post-rainy 2023 (without MC)	4.13	4.12	18.27	18.23	0.90	0.90
Post-rainy 2023 (with MC)	4.86	4.90	21.51	21.67	0.92	0.93
Post-rainy 2024 (with MC)	4.51	4.62	19.95	20.43	1.02	1.04
Gain (%) (without-with MC)	**15.02**	**15.92**	**15.06**	**15.87**	**2.17**	**3.23**
Gain (%) (with MC)[b]	**−7.76**	**−6.06**	**−7.82**	**−6.07**	**9.80**	**10.58**
Dry 2023 (with MC)	4.79	4.83	21.20	21.38	0.92	0.92
Dry 2024 (with MC)	4.78	4.95	21.15	21.92	0.89	0.92
Gain (%) (with MC)[c]	**−0.21**	**2.42**	**−0.24**	**2.46**	**−3.37**	**0.00**

[a] Pre-Rainy period 2024–2025 (with MC) taken as a reference for calculating the gain (%)
[b] Post-Rainy period 2024 (with MC) taken as a reference for calculating the gain (%)
[c] Dry period 2024 (with MC) taken as a reference for calculating the gain (%)

average cleaning interval increase from 17.64 days to 25.05 days. Conversely, the clean modules (ST2) showed gains of 2.42% in $Yield_{ST2}$ and 2.46% in CF_{ST2}, demonstrating the effectiveness of regular cleanings in mitigating soiling effects, even with the increase of the average interval between cleanings; PR_{ST2} remained stable.

Chapter 6
General Considerations

6.1 Summarizing

PV power increasing adoption worldwide has stimulated the search for strategies to optimize system performance. In this context, soiling effects on electricity generation and module temperature, along with mitigation approaches, have received growing attention in the literature. The analysis of the manual cleaning consequences on LEA2 performance highlights the influence of anthropogenic factors, such as repair works in a road, on soiling increase: $\%E_{LEA2}$ increased from 0.32% (1 day before MC3) to 2.79% (1 day after MC3). This behavior demonstrates the need for careful assessment of sites with frequent pollutant and dust emissions prior to PV plant installation.

During the rainy and post-rainy period of 2024, MC17 was followed by a $\%E_{LEA2}$ increase from 0.73% to 3.56%. This upward trend, despite frequent rainfall, indicates that natural cleaning alone is insufficient to fully recover soiled modules performance. Cleaning intervals should be planned based on seasonal precipitations, as heavy rainfall reduces, but does not eliminate, the need for cleaning of urban PV systems. In the driest period following MC26, $\%E_{LEA2}$ reached a maximum of 5.25%, confirming the intensification of soiling during such periods. Performance metrics analysis reinforces the effectiveness of regular cleaning procedures in maintaining LEA2 performance. During rainy periods, PR_{ST2} increase (5.21%) was 63.33% higher than that of P_{RST1} (3.19%), indicating that the combined effect of rainfall and manual cleaning enhances module performance.

Comparing the dry periods of 2023 and 2024, both with manual cleaning, all ST1 metrics showed a reduction. In contrast, clean modules showed gains of 2.42% in $Yield_{ST2}$ and 2.46% in CF_{ST2}, even with the increase of the average cleaning interval. These results emphasize that regular, seasonally adjusted cleaning schedules are

J. J. Silva de Souza, P. C. Marques de Carvalho, *Soiling Effects on Photovoltaic Systems*, SpringerBriefs in Applied Sciences and Technology,
https://doi.org/10.1007/978-3-032-22442-2_6

essential to minimize soiling losses in regions with low precipitation and high dust deposition potential. Summarizing, the results offer guidance for optimizing PV system operation in urban environments in semi-arid regions, in addition to providing the theoretical basis for developing seasonal maintenance strategies.

The manufacturer's authorised representative in the EU is Springer Nature Customer Service Centre GmbH, Europaplatz 3, 69115 Heidelberg, Germany. If you have any concerns regarding our products, please contact ProductSafety@springernature.com

Printed and bound by CPI Group (UK) Ltd, Croydon, CR0 4YY
07/07/2026
02160927-0003